T0138562

Eyes Turned Skyward; An Introduction to Aerospace Engineering with Emphasis on Aerodynamics and Aircraft Performance Analysis

by Patrick H. Browning and John L. Loth
Department of Mechanical and Aerospace Engineering
Benjamin M. Statler College of Engineering and Mineral Resources
West Virginia University
P.O. Box 6106, Morgantown WV 26506-6106

Printed by Morgantown Printing and Binding, Inc., 915 Green Bag Road, Morgantown, WV 26508

Distributed by West Virginia University Press, P.O. Box 6295, Morgantown, WV 26506

ISBN: 978-1-943665-02-0

This textbook is written for a one-semester introductory course in Aerospace Engineering. Although written at college level, designers and builders of experimental aircraft, UAV's and model aircraft can apply the content to design and estimate the performance of their aircraft. Theoretical derivations, which require knowledge of calculus, have been isolated in textboxes so that readers unfamiliar with calculus can skip over these textboxes and concentrate on the results in algebraic form.

WHY FLY? ONCE YOU HAVE TASTED FLIGHT,
YOU WILL FOREVER WALK THE EARTH WITH
YOUR EYES TURNED SKYWARD, FOR THERE YOU
HAVE BEEN, AND THERE YOU WILL ALWAYS
LONG TO RETURN.

- LEONARDO DA VINCI (1452-1519)

CONTENTS

FOREWORD

This text was written to introduce students to the thought process of an engineer on the subject of flight. This approach is to start descriptively, followed with basic principles, sprinkled with drawings, quantitative analysis and sample calculations for each topic. Drawing pictures on the back of an envelope is the way engineers communicate with one another and begin to solve problems. This text is the result of nearly 30 years of lecture notes for the West Virginia University course MAE 215 "Introduction to Aerospace Engineering". Its content has been modeled after two outstanding books on this subject. The first is "Introduction to Flight" by John D. Anderson, whose fifth edition covers a wide range of topics in 814 pages. The second is "Engineering Approach to Aerodynamics and Aircraft Performance" by John L. Loth. Loth's book has served as the primary model for this book, both in its structure and content, and without it this textbook would have been far longer coming to print. The current text concentrates on the fundamentals of fluid mechanics and thermodynamics needed to derive the three conservation equations and the equation of state. Added to these are the concepts of lift and drag to provide aerospace engineers with the tools needed to design aircraft and predict their performance. The derivations of the governing equations requiring calculus have been isolated in a bordered textbox. This facilitates skipping over them by readers not familiar with calculus. The resulting algebraic equations may be used in spreadsheet and engineering programming code, to calculate and predict the performance of subsonic aircraft, UAV's, and model aircraft.

This book has been assembled with the intention of making it as easy as possible for students to gain an appreciation for both the theoretical and applied aspects of modern aerospace engineering. Figures, tables, and critical equations are numbered and referenced throughout the book to support the written content as much as possible. Where it may help, example problems are provided that detail the entire engineering problem solving method as applied to the subject at hand. Additional problems are posed to the reader as "Test Your Understanding" cases in which only the initial problem statement and the final answer are provided; the process for arriving at the answer is left to the reader. When appropriate, small segments of engineering programming code have been placed into the text. Although written expressly for MATLAB®, the logic of the code is easy to follow and is thus easily modified to another programming compiler or language. Similar examples have been inserted demonstrating spreadsheet-based analysis using Microsoft® Excel®. Suggested homework problems are provided at the end of each chapter, while select answers to the chapter problems and additional problems are provided in the appendices. Section summaries have been provided for chapters where it seemed most appropriate. The last page of this textbook offers a handy reference table with important constants and helpful conversions based upon units common to aerospace engineering endeavors.

Based on the authors' experiences teaching this material for the last few decades, instructors using this text are highly encouraged to break the content of this text up in the following manner:

> Section 1 (basics of aerospace engineering, 3 weeks) – Chapter 1 through Chapter 3
> Section 2 (conservation laws and dimensionless parameters, 5 weeks) – Chapter 4
> Section 3 (wing and wing section aerodynamics, 3 weeks) – Chapter 5
> Section 4 (aircraft performance analysis, 2 weeks) – Chapter 6

Students and instructors alike will notice that this textbook has been designed to facilitate the teaching schedule suggested above.

HISTORY OF AVIATION AT WEST VIRGINIA UNIVERSITY

West Virginia University was established in 1867 as a Land Grant University with 184 students. Its Board of Regents established the School of Engineering in 1887. When in 1944 the WVU Department of Aeronautical Engineering was established it became one of only half a dozen such programs in the USA. More than 1000 students received their Private Pilot certification before courses in flight training were cancelled in 1966. One of the aircraft from this flight-training program, a 1966 Cessna 206-U was fully instrumented by the Naval Air Warfare Center for the senior Flight Testing class MAE 466. A 1959 Cessna 150 fuel- system was modified in 1995 to become the first alternate fuel (ethanol) aircraft with FAA certification for in-flight switch over between ethanol and aviation gasoline. As a result, John Russell, Director of DOE Alternate Fuels Utilization, designated the Morgantown Airport as the 2nd U.S. "Clean Airport" on August 28, 1996. A WVU research program on one of the most efficient high lift systems by blowing, called circulation control (CC) resulted in the design, construction and flight testing of the "CC Technology Demonstrator STOL Aircraft", which had its first flight April 10, 1974. A brief movie is available on YouTube entitled "WVU Circulation Control Aircraft" of this amazing aircraft in which its takeoff, flight, and landing characteristics, as well as the basics of its design, are all discussed and demonstrated. This technology was later installed and flown in 1979 on a Navy Grumman A-6A Intruder all weather bomber. Although the original WVU design used a variable geometry wing to produce high lift at the appropriate flight speeds, the later A-6A version exhibited a fixed geometry wing which limited its ability to perform well at high speeds. Until recently WVU offered a non-engineering credit class called Aviation Ground School for Private Pilots to encourage students to take up flying.

ACKNOWLEDGEMENTS

The satisfaction of observing students gain confidence by mastering the material in our classes (ranging from Introduction to Mechanical and Aerospace Engineering Design, Introduction to Aerospace Engineering, Compressible Aerodynamics, VSTOL Flight, Flight Vehicle Propulsion, Hypersonic Flight, Flight Testing, Aircraft/Spacecraft Design, AIAA Design-Build-Fly, etc.) has been most rewarding. Without students' continuous feedback and copies of their class notes, it would be impossible to sense their need in a textbook. When they graduate, most of them enter the work force with the feeling of having understood and remembered the material and an ability to apply these theories to real life problems. Invaluable online resources for some of the material presented in this book include the Bain Collection at the US Library of Congress, the UIUC Applied Aerodynamics Group website, and several excellent free information sites for XFLR-5 tutorials. Special thanks are given to the authors' colleagues, John Kuhlman, Gary Morris, and Frank and Eric Loth. Thanks also to colleagues Wade Huebsch and Pete Gall for their remarks on early notes from which much of this text is derived. Additional thanks are due to former students Kevin Ford and Will Vogel for providing early material and offering editing assistance. Robbie and Molly Browning deserve extra thanks for their expert assistance in designing the cover art for this textbook (if the reader finds the cover displeasing, it's entirely their fault).

SYMBOLS

a	lapse rate; speed of sound; three-dimensional lift curve slope; semi-major axis
A	area
b	wingspan; semi-minor axis
c	wing section and wing chord length; specific coefficient; mean random velocity
C	coefficient
d	diameter
D	aerodynamic drag
e	Oswald's spanwise loading efficiency factor; specific internal energy; eccentricity
E	endurance
F	force
g	gravitational acceleration
G	gravitational constant
h	altitude; height; specific enthalpy
i	electrical current
k	polytropic expansion; Boltzmann constant; parabolic drag polar coefficient
l	length
L	aerodynamic lift
L/D	lift-to-drag ratio (also known as glide ratio or finesse)
m	mass
M	moment
$\hat{n}$	unit normal vector
p	pressure
P	power
q	dynamic pressure
Q	heat flux
r	radius
R	gas constant; range
Re	Reynolds number
s	distance; curvilinear distance
S	reference area
t	time
T	thrust
u	specific internal energy; x-component of velocity
U	internal energy
v	y-component of velocity
V	velocity; voltage
$\forall$	volume
w	specific work; z-component of velocity; downwash velocity
W	weight; work
x, y, z	rectangular coordinates; distances
α	angle of attack
γ	ratio of specific heats; glide angle; climb angle
Γ	circulation vortex strength
δ	boundary layer height
Δ	incremental change
η	efficiency
λ	taper ratio
μ	dynamic viscosity
ν	kinematic viscosity
ρ	density
σ	Stefan-Boltzmann constant
τ	shear stress

SUBSCRIPTS

0	stagnation region; two-dimensional condition; at sea level
$1, 2, 3,...$	location 1, 2, 3,...
a, abs	absolute
air	air
amb	ambient
atm	atmosphere; atmospheric
B	buoyancy
c	chord length; centripetal
$c/4$	quarter-chord location
d	two-dimensional drag
D	three-dimensional drag
e	Earth
E	endurance
f	force; friction; final
$flight$	flight-direction
g	gauge (or gage)
G	geometric
h	horizontal tail surface; hydraulic
hw	headwind
i	incidence; initial; induced
l	two-dimensional lift
L	three-dimensional lift; landing
LO	lift-off
lam	laminar
$L=0$	zero lift condition
m	mass; mechanical
M	pitching moment axis
max, min	maximum, minimum
p	pressure-based; constant pressure; propeller
$para$	parasitic
r	wing root
$roll$	rolling
ref	reference
rev	reversible
sl	sea level
std	standard
t	wing tip
T	temperature-based
TO	take-off
$trans$	transitional
$turb$	turbulent
u	universal
v	vertical tail surface; constant volume
w	wing; wall
ρ	density-based
∞	infinitely far upstream
$\perp$	perpendicular

ACRONYMS AND ABBREVIATIONS

AIAA	American Institute of Aeronautics and Astronautics
AR	aspect ratio
ARDC	Air Research and Development Command
ATC	Air Traffic Control
AU	astronomical units
BL	boundary layer
BLC	Boundary Layer Control
BTU	British thermal unit
CC	Circulation Control
CS	control surface
C∀	control volume
FAA	Federal Aviation Administration
IC	internal combustion
ISR	Intelligence, Surveillance, and Reconnaissance
ke	specific kinetic energy
KE	kinetic energy
LE	leading edge
m.a.c.	mean aerodynamic chord
MAE	Mechanical and Aerospace Engineering
NACA	National Advisory Committee for Aeronautics
NASA	National Aeronautics and Space Administration
pe	specific potential energy
PE	potential energy
R/C	radio controlled; rate of climb
s.f.c.	thrust specific fuel consumption
S.G.	specific gravity
SI	Systeme Internationale
STOL	Short Take-off or Landing
UAS	Unmanned Aerial Systems
UAV	Unmanned Aerial Vehicles
UIUC	University of Illinois at Urbana-Champaign
WVU	West Virginia University

CHAPTER 1. INTRODUCTION TO AEROSPACE ENGINEERING

1.1 INTRODUCTION

A thorough understanding of the theory and practice of low speed flight is essential for the study of all flying objects ranging from fruit flies to supersonic aircraft and space shuttles. All of these must produce lift with minimal drag and be controllable at all speeds including those during take-off and landing. This material includes an introduction to compressible flow but limits all applications to low speed incompressible flow.

The study of the function of controls and instrumentation on some light aircraft as can be found at the local airport is essential to develop an appreciation for the requirements for flight. In addition constructing a small glider and test flying is instructive in appreciating the importance of the tail surfaces, the wing surface finish and the wing position relative to the center of gravity, for a successful flight. By the end of Chapter 5, students will be ready to design a glider model and analyze its performance on an Excel spreadsheet. By the end of Chapter 6 they will be ready to design a propeller driven model aircraft. Prior to construction it is recommended to have their designs reviewed by an experienced engineer.

To go faster than walking, one could ride a horse drawn cart. Such a single horse power (hp) engine only needs grass for fuel and repairs itself if damaged. To go even faster, one can take the train. Modern trains are pulled by 100,000 pound locomotives equipped with a 3000 hp diesel engine that generates electricity for the heavy duty electric motors at its hubs. The locomotive engine power, however, is small in comparison to the power produced by the four engines of a Boeing 747 on take-off. These four turbofan engines together produce about 200,000 pounds of thrust, at more than 100 mph, which is about 51,000 hp of thrust power. This is 17 times more power than that produced by the 3000 hp diesel locomotive. Even more amazing is the fact that these four aircraft engines together weigh only about one quarter of the locomotive engine's weight.

Most people enjoy fast and economical transportation on a commercial flight over long distances. An automobile is usually the fastest way of getting from one location to another within a one hundred-mile distance. For longer distances, airplane travel should be considered, as it is likely to be faster, more economical and safer than automobile travel. Consider an average 3500 pound automobile, carrying 2 passengers, at 58 miles per hour. Assume the car consumes one gallon of fuel for every 32 miles distance. The fuel economy for such a vehicle can be expressed as 2 x 32 = 64 passenger-miles-per-gallon. Next consider a Boeing 747 traveling at 580 miles per hour, or 10 times as fast as the automobile. When carrying 350 passengers and burning 3200 gallons per hour, its fuel economy is (580 mph/3200 gph) x 350 passengers = 64 passenger-miles per gallon. This means the fuel economy of the airplane is about the same as that of the automobile, but in an airplane one travels 10 times faster! The Boeing 747 can fly at least 20 times farther on the fuel in its tanks than an automobile can drive and can safely cover more than one million miles in between engine overhauls. A typical light airplane such as a Cessna 150, carrying two passengers at 90 mph burns 4.5 gallons per hour and will thus get (90 mph/4.5 gph) x 2 passengers = 40 passenger-miles per gallon. The Cessna 150 weighs only 1500 pounds, which is less than half as much as most automobiles. This is possible because airplanes do not need to be designed as collision resistant as an automobile.

1

EXAMPLE 1.1: EFFICIENCY OF A NEW LARGE AIRCRAFT

Given:

A newly proposed large airplane is expected to have increased fuel efficiency over older models. It is estimated to fly 9000 miles with 620 passengers while consuming 500,000 lb_f of Jet-A fuel (aviation kerosene). Jet-A fuel weighs 6 lb_f per U.S. gallon and costs $5.33 per gallon. Jet-A has a "lower heating value" of 18,400 BTU/lb of fuel. It is also known that 1 BTU (British Thermal Unit)= 1.055 kJ or 1 $kWhr$ = 3600/1.055 = 3413 BTU.

Find:

a) Calculate the volume of fuel consumed by such a new airplane (in U.S. gallons) and its efficiency (in passenger-miles per gallon) Note: A Boeing 747-400 averages 64 passenger-miles per gallon.
b) Calculate the fuel cost (in $) and the energy used for a long-haul flight (in $kWhr$).

Solution:

a)

$$V_{fuel} = \left(\frac{500,000\,lb_f}{}\right)\left(\frac{U.S.gal}{6\,lb_f}\right) = \boxed{83,300\,U.S.gal}$$

$$\text{efficiency} = \left(\frac{620\,pass}{}\right)\left(\frac{9,000\,mi.}{83,300\,U.S.gal}\right) = \boxed{67\,pass\cdot mi/gal}$$

b)

$$\text{cost} = \left(\frac{83,300\,U.S.gal}{}\right)\left(\frac{\$5.33}{U.S.gal}\right) = \boxed{\$444,000}$$

$$\text{energy used} = \left(\frac{500,000\,lb_f}{}\right)\left(\frac{18,400\,BTU}{lb_f}\right)\left(\frac{kWhr}{3413\,BTU}\right) = \boxed{2.70\times10^6\,kWhr}$$

Check:

The numerical values and the units appear to be reasonable and address the original question(s).

Seeing the ease with which birds can travel over obstacles such as mountains and lakes, it is no wonder that early man dreamed about flying. Actually all flying insects, birds and airplanes operate under the same laws of aerodynamics as used by aerospace engineers to calculate aircraft flight performance. Aerospace engineers can calculate the optimum flight speed and associated fuel consumption rate as a function of aircraft weight and wing area. Advantages of space flight are obvious to anyone who enjoys the benefits of communication satellites and cell phones. Their signals travel at the speed of light to bring us news, weather reports, and the world-wide-web. Flying is among the most fuel-efficient, and certainly the fastest, modes of transportation. The aircraft and space flight industry can therefore be expected to expand indefinitely.

To design a new airplane requires a large team effort by engineers of different disciplines. Only a small fraction of this team needs to be aerospace engineers. First the customer defines the number of passengers, weight of the cargo and range of the proposed aircraft. Based on such information, an aerospace engineer is needed to perform a preliminary analysis of different aircraft types capable of accomplishing the mission. This includes aircraft weight and fuel load, take-off and landing field distances required, noise aspects, cruising speed and altitude, safety, initial cost, operational cost, maintenance cost and life cycle. Based on this information the customer prepares an RFP (request for proposals) with the aircraft performance specifications, and the number of aircraft to be ordered. Such information is needed by aircraft manufacturers to prepare a competitive bid. The delivery price of the aircraft can be attributed to three distinct parts of the airplane, each of which cost about the same.

1. **Airframe:** The structure comprising of a fuselage to carry the payload, wings to generate lift to support the airplane weight and to have enough volume to store the fuel, wing ailerons near the wing tips for roll control during turns, wing flaps for slow flight during take-off and landing. A canard or horizontal tail surface with an elevator is usually used for nose up or down pitch control, and a rudder on the vertical tail surface for yaw control. A landing gear is also required.
2. **Propulsion:** Engines with propellers, fans or jet nozzles are required to generate the thrust required for acceleration and climb. This thrust level is more than the cruise drag.
3. **Avionics:** Avionics and control systems, including an autopilot with hydraulic or electrical control surface actuators. From the cockpit the pilot has control over all aspects of flight, with the exception of his external pre-flight inspection.

Needless to say, a large number of engineers of different specialization are required to prepare an accurate bid on such a new aircraft. Aerodynamic engineers determine the optimum wing design to produce the required lift at take-off and landing speed and minimize drag in cruise. They also design the streamlined fuselage and engine inlets. Flight control engineers design the control surfaces to provide static and dynamic stability and control over the entire operational envelope. Structural and material engineers are needed to design the lightest possible structure, which is usually a combination of a shell structure with spars and stringers. To prevent excess weight and/or over design, all parts are stressed to a safe limit based on a specified fatigue life. Electro-Mechanical (Mechatronic) engineers and production engineers design the numerous actuator mechanisms and linkages between the cockpit and the controls. Hydraulic engineers design the complex fuel management and supply line system as well as the hydraulic actuators for the flight control surfaces, landing gear, etc. Combustion engineers are involved in the jet engine design where the heat release rate per unit volume is typically 12,000 BTU per cubic foot as compared to 8 BTU per cubic foot in a conventional steam power plant, (BTU is British Thermal Unit). Electronic engineers are needed to develop the avionics required to navigate, communicate with ATC (Aircraft Traffic Control) and provide autopilot control from take-off to landing. Instrumentation engineers are needed for cockpit layout to allow the pilot to control and monitor all important flight parameters.

To become familiar with some major components of an airplane, some images are shown of the first series of Cessna 150 aircraft built (Figure 1-1 and Figure 1-2). Over 22,500 of this high wing, two seat training aircraft were built from 1959 to 1977.

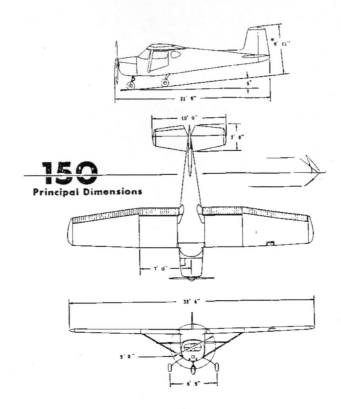

Figure 1-1: Cessna 150 drawing views: side, top, and front.

Figure 1-2: Cessna 150.

A conventional aircraft is made up of the following components:

1. The fuselage contains the cockpit, controls and the cargo. It is needed to connect all other components

2. The wing of reference area, S_w, provides lift, L, by reducing air pressure on the upper surface. In level flight the vertical upward lift compensates for the airplane weight, W.

3. The engine driven propeller provides the thrust, T, which in steady level flight compensates for the drag, D.

4. The vertical and horizontal tail assembly is called the empennage and in steady flight its lift is needed to cancel any pitching/yawing moments about the airplane center of gravity.

5. Wing ailerons are needed for roll control and to initiate and terminate a turn.

6. The landing gear is needed for efficient take-off and landing and to provide propeller ground clearance

1.2   HISTORICAL DEVELOPMENTS

The airplane did not become a reality until the 20th century. Why did it take so long for something so desirable? Cars and engines had been around a long time before an airplane could fly. The delay was essentially caused by the difficulties associated with flying. A brief list of some of the major challenges associated with flight is as follows:

- An airplane structure must not only be strong but also be light, because the thrust, T, required for steady flight is a function of its weight, W.
- Only if an airplane has an efficient wing can the required thrust, T, be reduced to the range from 10 % to 50% of the airplane weight, W. Note: Helicopters and rockets do not have wings and must be able to go straight up in the air, and therefore need thrust, T, greater than their weight, W.
- It was not until the 20th century that a propeller was developed capable of providing thrust in excess of the power producing engine weight.
- The weight distribution in a car is not critical, as a car is stable on four wheels. In an airplane, weight distribution is as critical as that of a unicycle, where the rider's weight vector must pass through the wheel-to-ground contact point. In a hang glider the pilot's weight is used for pitch and roll control. In level flight it must line up with the lift force on the wing. An airplane is too heavy to be balanced by the pilot's body weight alone. Thus three separate control surfaces are added to balance all moments in roll, pitch and yaw.
- All early design concepts were based on man-powered flight. As a top athlete cannot produce more than 1 HP, it is no wonder that man-powered flight did not succeed till the 1970's. A man-powered aircraft has to be extremely efficient in all aspects of the design. This was not possible until Paul MacCready invented the Gossamer Condor, and his team won the Kramer prize. This airplane is now on display in the Smithsonian Museum. Leonardo da Vinci, who by the end of the fifteenth century had made more than 500 sketches of proposed man-powered flying machines, had little chance of success.

The first two people to be lifted into the air were in a hot-air balloon that drifted a distance of 5 miles over Paris, France on November 21, 1783. The balloon (Figure 1-3) was designed and constructed by the Montgolfier brothers, Joseph and Etienne.

Figure 1-3: First aerial voyage in history: the Montgolfier hot-air balloon lifts from the ground near Paris, November 21, 1783 (National Air and Space Museum).

The Montgolfier's balloon created lift by heating the air inside with a fire that burned in the wicker basket suspended below. A hot air balloon is naturally stable by the pendulum principle because the payload basket hangs below the center of buoyancy of the balloon. Preliminary flight tests were done without human cargo but with a duck, rooster and a sheep on board.

After many investigators had failed to provide both lift and thrust using flapping bird-like wings, Sir George Cayley conceived the fixed wing modern airplane concept in 1799. Cayley separated the components needed for lift, thrust, pitch and yaw control. He proposed a fixed wing for lift; this wing was to be cambered for increased efficiency with a higher lift to drag ratio, as compared to a flat surface. He discovered this by aerodynamic testing of an airfoil mounted to a whirling arm, as shown in Figure 1-4.

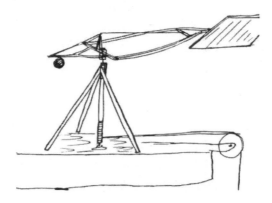

Figure 1-4: George Cayley's whirling arm apparatus for testing airfoils (National Air and Space Museum).

Rowing paddles were suggested for thrust, a vertical tail like a fish for yaw control and a horizontal tail like a whale for pitch control. Cayley tested the concept in 1804 with a meter long simple glider that performs and looks much like a modern glider. Etched into a small piece of metal, Cayley sketched out the first recognizable conventional aircraft design (Figure 1-5).

Figure 1-5: The first modern configuration airplane in history: Cayley's model glider, 1804 (National Air and Space Museum).

In 1809-1810, Cayley published a triple paper "On Aerial Navigation" in the Journal of Natural Philosophy. In 1849, he designed and tested a tri-wing glider, which carried a boy. On September 25, 1852, when he was 79 years old, he published in "Mechanics Magazine" the article "Sir George Cayley's Governable Parachutes" (Figure 1-6). This paper went unnoticed until 1960, otherwise some might argue that the airplane may have been invented a whole century earlier. Based on Cayley's designs, William S. Henson made a series of engravings of an aerial steam carriage (Figure 1-7). In 1842-1843 these prints were sold worldwide and stimulated work on an actual airplane.

𝔐echanics' 𝔐agazine,

MUSEUM, REGISTER, JOURNAL, AND GAZETTE.

No. 1520.]   SATURDAY, SEPTEMBER 25, 1852.   [Price 3d., Stamped 4d.

Edited by J. C. Robertson, 166, Fleet-street.

SIR GEORGE CAYLEY'S GOVERNABLE PARACHUTES.

Fig. 2.

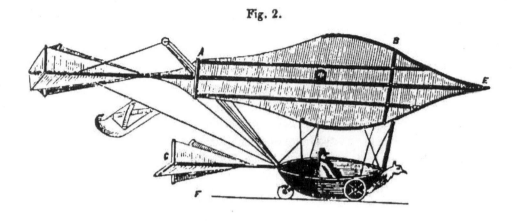

Fig. 1.

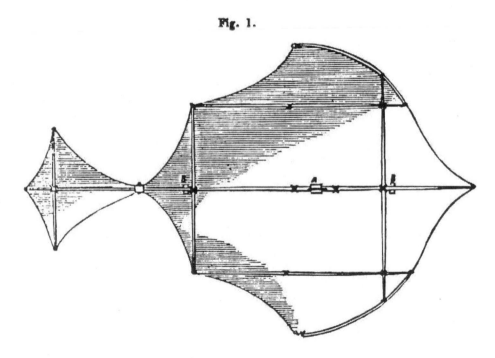

Figure 1-6: Sir Cayley's Governable Parachute (National Air and Space Museum).

Figure 1-7: Henson's aerial steam carriage, 1842-1843 (National Air and Space Museum).

In 1857-1878, the French naval officer Felix Du Temple flew the first successful powered model airplane in the world. Soon thereafter several piloted planes, launched from sloping chutes, managed to jump in the air for a moment, but did not achieve controlled flight.

C. H. Gibbs-Smith states the following criteria for successful powered flight: "To qualify for having made a simple powered and sustained flight, the airplane should have sustained itself freely in a horizontal or rising flight path – without loss of airspeed – beyond a point where it could be influenced by any momentum built up before it left the ground; otherwise its performance is rated as a powered leap! ... Maintaining adequate equilibrium in flight is part and parcel of sustention."

A daring Prussian from the late 19th Century by the name of Otto Lilienthal studied the structure and flight mechanics of a variety of birds and drew inspiration from the designs to pursue the construction of bird-like gliders large enough to carry a human adult. He personally flew these gliders (Figure 1-8) to scientifically study control schemes and made more than 2000 flights before dying due to injuries sustained in a crash. On his deathbed, Otto was reported to have told his brother, "Opfer müssen gebracht werden" (the German idiom means "Sacrifices must be made").

Figure 1-8: A monoplane glider by Lilienthal, 1894 (National Air and Space Museum).

Lilienthal's student, Percy Pilcher, was killed in 1899 in a crash of his glider, the "Hawk", which he intended to fly as an airplane with a 4HP motor and a 5 ft. diameter propeller.

Samuel P. Langley was appointed in 1887 as secretary of the Smithsonian Institution. In 1898, he was given $50,000 – a huge sum of money in those days – by President William McKinley to build a machine to carry passengers. Although his small scale models had been successful, Langley's full scale "Aerodromes" failed. His second effort failed on December 8, 1903, just nine days prior to Wilbur and Orville Wright's first successful airplane, the "Wright Flyer I" (Figure 1-9), made its historic flight on December 17, 1903 at Kill Devil Hills, NC. Orville Wright took off from a 60 ft. horizontal launching rail. After unevenly rising to about ten feet, the Flyer darted back into the sand, a distance of 120 feet from the launching rail.

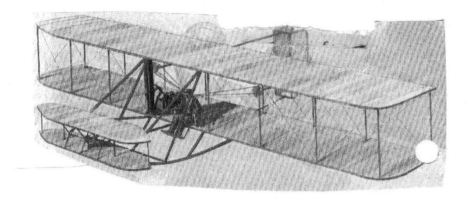

Figure 1-9: The Wright brothers' first heavier-than-air flight in history (National Air and Space Museum).

Three more flights were made that morning, with the final one lasting 59 seconds while covering a distance of 852 feet. The Wright Flyer I used two 8.5 ft diameter propellers, driven by one 12 hp motor weighing 280 lb. The Wright brothers' success was due to prior careful experimental testing of more than 200 airfoil shapes in their home-built wind tunnel, followed by three years of trying the best airfoils on the wings and control surfaces of their gliders. After their public demonstration flights in the U.S. and Europe in 1908, the development of aeronautics took off exponentially. In 1926, the Ford Motor Co. introduced the highly successful Ford 4-AT Trimotor. General Motors maintained its airplane engine division, the Allison Division in Indianapolis, until being sold to Rolls-Royce in 1995. Today a wide variety of automobile engines are in use in home built aircraft.

Progress in propulsion engines has been driven by the need to fly faster and higher. During World War II radial piston engines with four rows of seven radially-aligned cylinders each were developed which produced as much as 2200 hp. Both in England and in Germany, separate inventors developed the jet engine and the rocket motor for aircraft propulsion. Both these new engines flew successfully in the skies over Germany towards the end of the war.

Currently, GE jet engines that are more than 10 feet in diameter and produce 100,000 pounds of thrust are used in the Boeing 777. They each produce up to 24,000 hp – equivalent to the power of eight 3000HP diesel locomotives at a railroad yard.

The space shuttle solid booster rocket motors produce 2,700,000 pounds thrust. After 120 seconds they burn out at a flight speed of about 5000 ft/sec. At that time they produce 24.8 million hp. Liquid hydrogen and oxygen rocket motors are then needed to accelerate the space shuttle to its orbital velocity of almost 26,000 ft/sec.

CHAPTER 1 PROBLEMS

1.1) A newly proposed large airplane is expected to be more fuel efficient and fly 9000 miles with 620 passengers, or 620 x 9000 passenger-miles, while consuming 500,000 lb_f of Jet-A fuel (aviation kerosene). Given is: Jet-A fuel weighs 6 lb_f per U.S. gallon and costs $2.50 per gallon. Jet-A has a "lower heating value" of 18,400 BTU/lb_m of fuel. Note the following energy conversions: 1 BTU (British Thermal Unit)= 1.055 kJ or 1 kWhr = 3600/1.055 = 3413 BTU.

    a. Calculate the fuel consumed by such a new airplane (in gallons), and fuel-efficiency (in units of passenger-miles/gallon). Compare your answer to that of the Boeing 747-400 which averages 64 (passenger-miles/gallon). (Note: "compare" means calculate the % increase or decrease)

    b. Calculate the fuel cost in $. Calculate energy used in kWhr and how much this mission would have cost if the engines were powered by electricity costing $0.06 per kWhr. If electricity turns out to be cheaper, why then not use it instead of fuel?

1.2) Describe the function of each airplane component: wing, aileron, flap, fuselage, empennage and propeller.

1.3) Fill in the blank: To fly in steady level flight the pilot must actuate the controls and engine throttle such that the forces and moments about the airplane's center of gravity are in _____.

1.4) Who made the first aerial voyage? In what year was this accomplished?

1.5) Who was the person to design the first concept of the modern airplane configuration, separating the component for lift and thrust generation into a fixed wing for lift, a separate engine for thrust to counter the drag and to further employ separate horizontal and vertical surfaces in the tail for pitch and yaw control?

1.6) What kind of modern aircraft generates both lift and thrust with one component?

1.7) The jet engine was first developed during World War II for aircraft jet propulsion. Today it is used in a variety of other applications. List at least three. (Note that a jet must be supplied with air to facilitate the combustion of fuel, thus it must "breathe". Although some hybrid rockets employ breathing tubes, rockets as a rule do not need to "breathe" since they have their air/fuel mix stored on-board. In other words, rockets are NOT jets!)

CHAPTER 2.   UNITS AND EQUATION OF STATE

2.1   INTRODUCTION

Aerospace Engineers are trained in all aspects of vehicle design for travel away from the Earth's surface. This means travel through water, air or space. To design such a vehicle involves analysis of the fluid forces acting on all its surfaces, especially those surfaces providing lift and control. These fluid forces are needed to calculate the control surface areas needed to navigate the vehicle. They are also needed to calculate the propulsive thrust required for climb and acceleration in addition to that associated with canceling vehicle drag. The vehicle structure must be designed to cope with both static and dynamic fluid and inertial forces.

To calculate Fluid forces on a vehicle and the associated propulsive thrust requires knowledge of five aerodynamic variables as a function of space and time (x, y, z, t):

1.   *Velocity, V*
2.   *Pressure, p*
3.   *Temperature, T*
4.   *Density, ρ*
5.   *Shear stress, τ*

To solve for these five variables, the aerospace engineer uses five equations:

1.   *Continuity Equation* (conservation of mass, to find *V*)
2.   *Momentum Equation* (conservation of momentum for inviscid flow, to find *p*)
3.   *Energy Equation* (conservation of energy, used to find *T* from *V*)
4.   *Equation of State* (used to find *ρ* from *p* and *T*)
5.   *Navier-Stokes Equation* (momentum equation in viscous flow regime with an appropriate turbulence model for shear stress, τ, or empirical drag data)

These five equations show up in numerous different forms and levels of complexity in classes on aerodynamics ranging from subsonic to supersonic and hypersonic flow, aircraft and rocket propulsion, stability and control, aircraft design and flight-test classes.

1.   Viscous effects are often negligible, in which case the number of variables and number of equations required to solve for them reduces to four.
2.   For low speed aircraft, e.g., at flight speeds below 300 ft/sec (205 mph), the air density changes by less than 5%, thus its change is often neglected. In those cases the density, ρ, and the temperature, T, are treated as known constants. This reduces the number of unknowns to three variables and eliminates the need for the Equation of State and the Energy Equation.
3.   At low velocity, aircraft thrust is most efficiently generated by the use of a propeller (or a rotor in the case of a helicopter). The airflow around an aircraft and through the propeller produces all of the aerodynamic forces of interest, such as lift, L, thrust, T, and drag, D.
4.   The wing shape is designed to produce enough lift, L, to support aircraft weight, W, in level flight at minimal drag. Note sufficient lift can only be produced if the aircraft velocity exceeds the stall-speed, V_{stall}.

5.   Propeller blades are shaped in the form of airfoils. The forward component of the lift force produces the propeller blade thrust, which increases with RPM.

6.   Aircraft control surfaces are in the shape of airfoils to generate asymmetric lift forces, as needed for control of pitch, roll and yaw about the aircraft center of gravity.

7.   All aircraft components are streamlined (aerodynamically shaped) to minimize drag. The study of airflow, in and around aircraft components is called "aerodynamics", and persons practicing this science are aerodynamicists.

2.2   UNITS

In a given situation, an engineer must often be able to quantify one or more variables in order to fully describe its current state (or make predictions about past or future states). An extensive variety of units of measure must be used to quantify these variables. Fundamentally, variables can usually always be described in terms of either force-length-time (F-L-t) or mass-length-time (m-L-t) (note that one must add an additional "T" for temperature if they are working a typical fluid or thermodynamics problem).

Whenever groups of people must engage in some type of measurement (e.g., measurement of a bushel of wheat for market), an important condition immediately arises: there must be some common method of quantifying information so that accurate exchange can occur. Engineers must determine the system of units that they intend to use to accurately account for each variable in a problem, as well as to simplify the requirement to move between F-L-t and m-L-t units. In almost all branches of engineering, both the Systeme Internationale (SI, "metric") and the English (Imperial) systems are used. A brief list of some of the more common fundamental engineering units is given in Table 2-1.

Table 2-1: Some of the fundamental engineering units of the SI and English systems.

Variable	Symbol	SI	English
Force	F	newton (N)	pound (lb_f)
Length	L	meter (m)	feet (ft)
Time	t	second (sec, s)	second (sec, s)
Mass	m	kilogram (kg)	pound (lb_m) slug (slug)
Temperature	T	Celsius (°C) *Kelvin (K)	Fahrenheit (°F) *Rankine (°R)

*These are absolute scales indicating an inability for a gas to fall below zero Kelvin (0 degrees Rankine)

How does one move between F-L-t and m-L-t units? From Newton's second law, the force exerted on an object is equivalent to the change in the object's momentum. If the mass of the object is constant, then the force is the product of the object's mass and its resultant acceleration, or

$$F = ma$$

In metric or Systeme Internationale (SI) units this gives:

$$1\,N = 1\,kg \times 1\,m/\sec^2 \tag{2.1}$$

This system of units has a simple one-to-one unit relationship (i.e., one base unit of force for one base unit of mass for one base unit of acceleration). It forms the basis for the relationship between the four fundamental units: force, mass, length and time. It is also used to transform units. For example, acceleration can be written in units of (m/sec²) or in (N/kg), and pressure can be written in units of (N/m²) = pascal (Pa) or in (kg/m·sec²) because (N) = (kg·m/sec²).

Unfortunately names like "pounds" and "kilogram" are often used for both "force" and "mass" units, which creates confusion. For example, when someone enters a grocery store and asks for a pound of apples, they may not necessarily get a pound mass (lb_m) of apples, because the grocer uses a spring scale calibrated at sea-level in pounds force (lb_f) to fulfill the order. Only when the purchase is made at sea level, where the acceleration of gravity, g, equals the standard value (g_o = 32.2 ft/sec²), will one pound force of apples give the patron the same amount as one pound mass of apples. At 100,000 ft altitude, he would get 1% more apples. In the extreme case of gravity-free space it would take an infinite number of apples to weigh a pound (lb_f)! Similarly in Europe, all spring scales are calibrated erroneously in (kg) instead of (N) newtons of force. Europeans utilize the mass unit (kg) to indicate their weight, which is a force unit like the newton (N). The English System of units is also confusing. Consider the following relationship:

$$1\,lb_f = 1\,lb_m \times 32.2\,ft/\sec^2 = 1\,lb_m \times g_0 \tag{2.2}$$

Here, a factor g_0 = 32.2 ft/sec² is required to correlate force and mass units. This system of units does not have a simple one-to-one unit relationship like the metric SI system. Students and professionals alike are sometimes confused by the difference between equations published in English units, with g_0 in it, and those in metric units, without any reference to g_0. Throughout history, engineers have made costly unit conversion mistakes caused by this factor. Therefore in aerospace engineering courses the use of the unit pound mass (lb_m) is avoided. Instead the Engineering English System of units is used exclusively as illustrated in the following equation:

$$1\,lb_f = 1\,slug \times 1\,ft/\sec^2 \tag{2.3}$$

This equation has the same one-to-one unit relationship as the metric equivalent. It forms the basis for the relationship between four fundamental units: force, mass, length and time. For example, when used to transform units, acceleration can be written as (ft/sec²) or in (lb_f/slug), and pressure in (lb_f/ft²) or in (slug/ft·sec²), etc. In the event that English unit conversions involving both mass and force are needed, it is important to see that Eq. (2.3) may be rewritten to express unity, or

$$1 = \left(\frac{slug \cdot ft}{lb_f \cdot \sec^2} \right) = \left(\frac{lb_f \cdot \sec^2}{slug \cdot ft} \right) \tag{2.4}$$

Though perhaps not obvious, many equations utilizing English units of both mass and force may be successfully treated by carefully multiplying though by Eq. (2.4). Remember, it is always alright to multiply anything one wishes by one.

TEST YOUR UNDERSTANDING: CONVERTING ENGLISH FORCE AND MASS

The speed of sound, a, in English units is given in ft/sec and can be found as

$$a = \sqrt{\gamma RT}$$

For standard sea level air,

$$\gamma = 1.4 \qquad\qquad R = 1716 \, ft \cdot lb_f / slug \cdot °R \qquad\qquad T = 519 °R$$

At first glance, the product of γ, R, and T will yield units of ft·lb$_f$/slug. It is certainly not obvious that the square root of those units will indeed yield units of ft/sec.

Use Eq. (2.4) to show how the units for a are in fact ft/sec, and determine the value of a at sea level.

Answer: $a = 1{,}117 \, ft/sec$

An aerospace engineer must get used to thinking in (slugs) instead of (lb$_m$). Most important is that equations written in metric or SI units are then identical to those written in English units. In this text only metric and engineering English units will be used:

1. Length in meters (m) or feet (ft)
2. Mass in kilograms (kg) or slugs (slug) (Note: 1 slug = 32.2 lb$_m$ or 32.2 "pounds mass")
3. Force in newtons (N) or pounds force (lb$_f$)
4. Time in seconds (s or sec)
5. Temperature in Kelvin (K), degrees Celsius, Fahrenheit, or Rankine (°C, °F, or °R, respectively)

One significant and highly useful component related to the use of SI units is that a collection of multiplicative prefixes are available to help indicate the magnitude of the value in question. For example, a distance of 1,220 meters (1,220 m) may just as well be expressed as 12.2 hectometers (12.2 hm) or 1.22 kilometers (1.22 km). Similarly, 2 millionths of a gram (2 g or 2 gm) may be expressed as 2,000 milligrams (2000 mg) or 2 micrograms (2 µg). Table 2-2 provides an overview of the various prefixes for use with SI units.

Table 2-2: Prefixes for use with the Systeme Internationale

Prefix	Symbol	Exponent	Prefix	Symbol	Exponent
yotta	Y	10^{24}	deci	d	10^{-1}
zetta	Z	10^{21}	centi	c	10^{-2}
exa	E	10^{18}	milli	m	10^{-3}
peta	P	10^{15}	micro	µ	10^{-6}
tera	T	10^{12}	nano	n	10^{-9}
giga	G	10^{9}	pico	p	10^{-12}
mega	M	10^{6}	femto	f	10^{-15}
kilo	k	10^{3}	atto	a	10^{-18}
hecto	h	10^{2}	zepto	z	10^{-21}
deka	da	10^{1}	yocto	y	10^{-24}

TEST YOUR UNDERSTANDING: USING SI PREFIXES IN AN ELECTRICAL CIRCUIT

Electrical power, P (in Watts, W), dissipated in a resistor can be found as the product of the current, i (in Amperes, A), through the resistor and its corresponding voltage drop, V (in Volts, V), such that

$$P = iV$$

If the power dissipation for a given resistor is known to be 500 mW, what is the current through it (in μA) if the voltage drop across it is 1.5 kV?

Answer: $i = 333\mu A$

The five fundamental aerodynamic variables are point-functions, meaning they are defined at a specified Point 1 (x_1, y_1, z_1) and may be different at some other Point 2 (x_2, y_2, z_2). These variables are:

1. Velocity, V, in units of m/sec, or ft/sec
2. Absolute pressure, p, in units of N/m² =Pa (pascal) or lb_f/ft² (psf)
3. Absolute temperature, T, in units of K or °R
4. Density, ρ, in units of kg/m³ or slug/ft³
5. Shear stress, τ, in units of N/m² (Pa) or lb_f/ft² (psf)

It may be easier for the reader to see these variables in a tabulated format such as Table 2-3:

Table 2-3: The five fundamental aerodynamic variables and their related units

Variable	Symbol	Fund. Units	SI	English
Velocity	V	Lt^{-1}	$(m/\sec)$	$(ft/\sec)$
Pressure	p	FL^{-2}	$(N/m^2),(Pa)$	$(lb_f/ft^2),(psf),(lb_f/in^2),(psi)$
Temperature	T	(T)	$(°C),(K)$	$(°F),(°R)$
Density	ρ	mL^{-3}	(kg/m^3)	$(slug/ft^3),(lb_m/ft^3)$
Shear	τ	FL^{-2}	$(N/m^2),(Pa)$	$(lb_f/ft^2),(psf),(lb_f/in^2),(psi)$

Many extremely handy conversions may be found on the last page of this textbook to help move between the values of important variables that present themselves in different unit systems.

2.3   PRESSURE IN ITS VARIOUS FORMS

INTRODUCTION

Fundamentally, the effect in gases which is known as pressure is the result of countless collisions during which momentum is exchanged between gas molecules and themselves or gas molecules and the walls by which they are contained. It is critical to understand that pressure always and only acts inward into a surface. If one imagines thousands of tiny molecules coming into contact with a nearby surface, there are only two ways to transfer momentum with the surface: the first is by the component of changing molecular velocity normal to the surface, and the second is by the component of changing velocity tangential to the surface. The first phenomenon is pressure, the second is shear stress. It should be obvious from this mental picture that the gas molecules are incapable of a "pulling" type of momentum exchange normal to the surface they interact with in an upward direction. The main concept to take away from this picture is that true, absolute pressure is always positive (or at the very least zero) and can never be negative, thus, there is no such thing as "suction" in an absolute sense. Rather, the concept of "suction" (i.e., that pressure which appears to pull on an object) only makes sense in a relative way. For example, the upper surface of a wing has relatively lower pressure in comparison to its corresponding lower surface, thus the upper surface flow appears to pull the wing up. Although pressure is a fundamentally simple concept, its influence can be expressed and utilized in a wide variety of ways by engineers and for this reason can sometimes be a bit confusing. In the following sections, an attempt will be made to explain some common "forms" for which pressure presents itself in engineering problems.

ABSOLUTE AND GAUGE PRESSURE

One of the defining characteristics of matter when it appears in the form of a gas is that it must take the form of the container within which it is held. It is often the case that gas molecules are contained in distinctly different containers for various reasons. Air is contained within car tires to provide a pneumatic cushion between the hard wheel and the hard road, thus providing the driver and passengers with far better driving comfort. Argon is commonly used as the inert shielding gas for welding, and comes in various sizes of high pressure steel tanks which fit neatly on or near the welding station where they are to be used. Helium that has been pumped into small rubber or Mylar balloons is often used at children's birthday parties to provide lighter-than-air floating decorations. The exact gas pressure in all of these cases is extremely important – improperly inflated car tires can result in poor fuel economy and in the worst cases death from vehicle loss of control events; a poorly pressurized argon tank can lead to fouled and weakened weld joints due to insufficient shield gas supply; everyone is familiar with the result when a balloon is over pressurized. Clearly, pressure is important, but exactly which pressure is being illustrated in these examples? There are in fact three distinctly different but interrelated pressures that may be applied to gases in containers:

1. absolute pressure, p_{abs} – the true pressure associated with the gas inside the container
2. ambient pressure, p_{amb} – the local pressure surrounding the outside of the container
3. gauge pressure, p_g – the difference between the absolute and the ambient pressure

These three pressures are related to one another according to a simple formula, given as

$$p_{abs} = p_g + p_{amb} \qquad (2.5)$$

When someone checks the pressure inside their car tire, they use a pressure gauge which inevitably has one of its ends exposed in some way to the local ambient pressure while the other is exposed to the air in the tire. This causes the device to actually read in terms of the difference between the pressure inside the tire and the pressure outside the tire, thus if the manufacturer recommends 35 psi be kept in the tire and the local pressure where the car is being driven is 14.5 psi, then the absolute pressure that is required to be in the tire is actually the sum of the gauge and ambient pressures, or 49.5 psi.

To eliminate confusion about which pressure is meant, the terms "absolute" or "gauge" are often added to pressure specifications to ensure the end user has properly filled whatever device he or she may be using. For car tires, gauge pressures are typically specified, so that the official pressure statement will read "Fill to 35 psig" where the "g" is short for "gauge". In the case of a pressure vessel for which someone is trying to compute the mass of gas within it, only absolute pressure can help determine this value. To compute the gas in the supposed car tire example, one would say that it has a pressure of "49.5 psia" where the "a" is short for "absolute". In almost all aerospace engineering applications, using absolute pressure when performing calculations is imperative so that one does not end up determining embarrassingly erroneous negative values of mass or volume.

One final note before moving to the next section: English is a living language and as such tends to experience changes in accepted spelling and usage as it evolves. It is almost certain that the reader has seen and perhaps used the word "gage" where the authors have used the word "gauge". It is the opinion of the authors that either of the two spellings is acceptable, and in the job of engineering problem solving, far greater problems exist than the question of which version of the word should be used. The careful observer may note and suppose that the subscript "g" in Eq. (2.5) is deliberate.

PRESSURE AS A COLUMN OF FLUID AND MANOMETERS

Much to the consternation of outsiders, aerospace engineers, particularly those involved in experimental aerodynamics, often give pressure in terms of lengths of fluid. Among the many ways to state pressure, it is quite correct to write

$$1\,atm = 2,116 \; lb_f \big/ ft^2 = 760\,mmHg$$

While it may not be immediately obvious, it is relatively easy to show why pressure which is fundamentally measured in terms of force per unit area can be specified in terms of a length or column of fluid. Consider a container of liquid mercury which is placed on a dining room table. Assume the bottom of this container measures exactly 1 m x 1 m. This case is shown pictorially in Figure 2-1. Now consider that the element mercury has a specific gravity (S.G.) of 13.6, indicating that it is 13.6 times denser than water, or

$$S.G._{Hg} = \frac{\rho_{Hg}}{\rho_{H_2O}} = 13.595$$

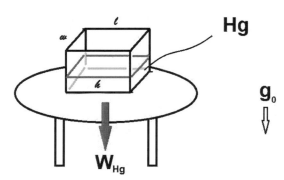

Figure 2-1: Liquid mercury is held in a tank with a base length and width of exactly 1 m.

Further, the weight of the mercury may be calculated as

$$W_{Hg} = m_{Hg}g_0 = \rho_{Hg}V_{Hg}g_0 = 13.595\rho_{H_2O}(l \times w \times h)g_0$$

so that

$$W_{Hg} = \frac{13,595\,kg}{m^3}\left|\frac{9.8067\,m}{sec^2}\right|\frac{N \cdot sec^2}{kg \cdot m}\left|\frac{1\,m^2}{}\right|\frac{h}{} = 133,322\,h\,N/m$$

If the container has constant width and length, then what dimension does the weight per unit area of table space of the mercury depend upon? The only variable that governs the total weight per unit area on the table is actually the height of the mercury. Now consider what the overall weight of the mercury would be if the mercury column, h, was exactly 760 mm deep:

$$W_{Hg} = 133,322(0.760\,m)\,N/m = 101,325\,N$$

Now calculate the weight per unit area that this container exerts on the surface of the table:

$$p_{Hg} = \frac{101,325\,N}{1\,m^2} = 101,325\,Pa = 1\,atm$$

If one recreates this example with a smaller or larger surface on the bottom but the same height, he or she will soon discover that although the weight of the fluid changes, the pressure (i.e., the weight per unit of table surface area) exerted on the surface remains at a constant value. In other words, a length of a fluid column can be used to describe force per unit area without the need for discussion of the size of the bottom area of the column.

One may begin to suspect that a device which contains a fluid volume that can freely move up and down within a tube might be able to successfully measure pressure. A U-tube manometer is a simply-constructed device made by making a U-shape out of a tube so that it can be filled with liquid with each of its two openings turned upright and exposed to different inlet pressures. Consider water in a U-tube manometer with one leg which has been connected to a pressure vessel while the other leg is open to the ambient pressure as shown in Figure 2-2.

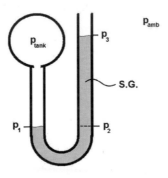

Figure 2-2: A U-tube manometer is filled with fluid and connected between ambient air and a pressure vessel filled with some compressed gas. Pressure in the tank is approximately the same as the pressure at locations 1 and 2, and the ambient pressure is approximately the same as that at location 3. The specific gravity (S.G.) of the fluid determines how far the column moves when there is a pressure difference between location 1 and location 3.

If one is able to measure the difference in height between locations 1 and 3 (call that *Δh*), and if one knows the specific gravity (S.G.) of the fluid in between locations 1 and 3, then it is possible to determine the difference in pressure between these two locations. Intuitively, one can guess that the pressure at location 1 is higher than the pressure at location 3 based solely on the fact that the column inside the U-tube has shifted to the side open to the atmosphere. Also note that the tank pressure is effectively the same as the pressure at location 1 and that the ambient pressure is effectively the same as the pressure at location 3. If the ambient pressure is known, then one may also be able to figure out the pressure in the tank. The pressure in the tank may be calculated as

$$p_{tank} = p_{amb} + \rho_{fluid} g \left(h_3 - h_1 \right) = p_{amb} + S.G. \rho_{H_2O} g \left(\Delta h \right)$$

In general, for manometers, one may write

$$p_1 = p_2 + \rho_{fluid} g \left(h_2 - h_1 \right) = p_2 + S.G._{fluid} \, \rho_{H_2O} g \left(h_2 - h_1 \right) \tag{2.6}$$

TEST YOUR UNDERSTANDING: CALCULATING PRESSURE USING A DRINKING STRAW

A child who has never heard of manometers has experimented with pressure and manometers without even realizing it. By inserting a normal drinking straw down into her glass of milk, she begins to gently increase the pressure in her mouth until bubbles begin to churn up from below the surface of the milk.

If the distance between the submerged end of the straw and the surface of the milk in the cup is exactly 3 inches, the ambient pressure is 14.7 psia, the local gravitational acceleration is 9.81 m/sec², and the specific gravity of milk is 1.032, what is the minimum absolute mouth pressure required to blow milk bubbles? Please provide your answer in psia and Pa.

Answer: $p_{mouth} = 14.8 \, psia = 102,096 \, Pa$

2.4   DENSITY

Density is simply a measure of the amount of mass of a substance held within a specified volume. Mathematically, this is expressed as

$$\rho = \frac{m}{\forall} \tag{2.7}$$

The density of a gas has a tremendous impact on a number of critical aerodynamic parameters. Mass flow rate, dynamic pressure, lift and drag – all of these parameters vary directly with density.

2.5   TEMPERATURE AND TEMPERATURE SCALES

A gas is an assembly of molecules, which move around the container at high random velocity, c. Their velocities change direction and magnitude upon collision with one another or with the walls of the container. The associated momentum exchange provides the pressure in the container. On average, the square of their molecular random motion velocity, c, is the sum of the squares of their *x*, *y*, *z* component velocities (*u*, *v*, *w*) such that

$$c^2 = u^2 + v^2 + w^2 = 3u^2$$

The internal random translational kinetic energy (KE) per molecule is $0.5(c^2) = (3/2)(u^2)$ in units of $(m/sec)^2/kg$ or in N·m/molecule = joule/molecule. Using the Boltzmann constant, k = 1.38 x 10^{-23} in units of Joules per molecule per Kelvin gives the random translational KE per molecule = $(3/2)kT$. Per kg mass of gas, this KE is $(3/2)RT$, where R is called the specific gas constant. Per kmol mass this KE is $(3/2)R_uT$, where R_u is called the universal gas constant. This equation shows that the molecular random motion kinetic energy is directly proportional to the absolute temperature, T, of the molecules. In outer space molecules move at high velocity, but no longer collide with one another, thus cannot exert force on each other or maintain their random KE. Then $T = 0$ K and therefore the absolute pressure in outer space is zero ($p = 0$ Pa = 0 N/m^2). The experimental discovery that a perfect gas sealed in a container increases in pressure in direct proportion to its temperature, provides a means of correlating common scales such as °C and °F to absolute scales such as K and °R.

Galileo Galilei (1564-1642) is credited with constructing the first practical thermometer. He used an arbitrary scale and called it a thermoscope. It was improved in 1650 by completely sealing the indicating liquid inside a tube. Daniel Fahrenheit defined °F in the year 1724, based on 0 °F as the predicted lowest temperature for an ice/salt/water mixture and an approximately 100 °F (actually 98.6 °F average) body temperature. Anders Celsius defined °C in the year 1742 based 0 °C for water freezing and 100 °C for water boiling at sea level on a standard day. This 100 °C boiling temperature is equal to 212 °F while 0°C freezing temperature is equal to 32°F. In the year 1848, William Thomson, who became Lord Kelvin developed the absolute temperature scale in K. This was shortly after William Rankine introduced °R. For reasons to be explained later, only the absolute temperature scale in either K or °R should be used in aerodynamics. Examples are equation of state, the definition of speed of sound, the Mach number and the maximum velocity obtainable by a gas.

Temperature, then, provides an indirect way of determining the mean random motion of a gas' molecules. While the molecules move around at random, they bounce off the walls of the container like basketballs. Upon each collision their velocity and momentum changes direction. This momentum exchange with the walls of the container exerts an average normal force $\Delta F_\perp$ on differential wall area ΔA. The average normal force per unit area $(\Delta F_\perp/\Delta A)$ is called the average pressure, p. When the area ΔA is reduced to the minimum size possible to maintain a pressure independent of its size, then this becomes the pressure at a point (x, y, z).

Perform the following experiment at sea level on a standard day where the barometer reads $p_1 = 760$ mm Hg or the pressure is 1 atm. Take a constant volume, V, container, filled with air. Submerge the container, which is open at the top, in a bucket filled with water, and bring the water to a boil. At this pressure of 1 atm boiling is defined to occur at a temperature $T_1 = 100$ °C or 212 °F. Seal off the container by connecting it to a mercury manometer. A manometer is basically a plastic tube with mercury inside, mounted in the shape of a "U" fastened to a vertical board. The level of the mercury in both legs of the "U" will be the same, as pressure is 1 atm inside and outside. After removing the heat source, begin taking p and T data while adding ice-cubes to the water bucket. When enough ice has been added so that it won't melt anymore, the temperature has dropped by definition to $T_2 = 0$°C or 32°F. At that time, one will notice that the pressure in the container has also dropped. The manometer legs at this moment will show a difference of 203 mmHg. The pressure in the container is now $p_2 = 760-203 = 557$ mmHg. If one keeps on cooling, the pressure will keep on dropping, theoretically all the way to zero absolute pressure, $p_3 = 0$. Plotting the temperature in °C on the horizontal axis, and the pressure data in mmHg on the vertical axis shows a linear behavior between pressure and temperature (Figure 2-3).

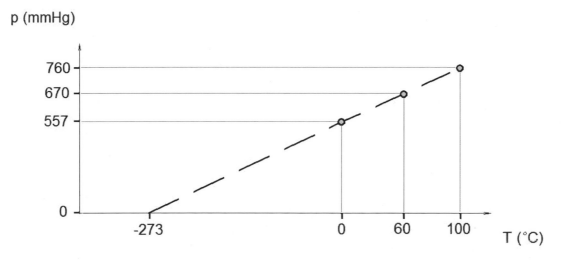

Figure 2-3: Plotting the change in pressure as a function of temperature. If one extrapolates to a pressure of 0 mmHg, the temperature must reduce to -273 °C or 0 K.

At p_3 = 0, molecules no longer bounce into each other, stopping random molecular motion. As random motion KE is a function of temperature, one has arrived at absolute zero, where T_3 = 0. To calculate the corresponding thermometer reading, extrapolate the data points in Fig. 2-1 to p_3 = 0 and read T_3 = -273 °C (or -460 °F). Notice that all three-test points are on a straight line. This indicates that the absolute pressure in a closed rigid vessel is only a linear function of its absolute temperature T in K. Note T_3 can also be computed from: $(T_1\text{-}T_3)/(T_2\text{-}T_3)$ = $(p_1\text{-}0)/(p_2\text{-}0)$.

Another technique could be used, which is to measure at constant pressure the change in volume of a friction-free piston-cylinder arrangement. One will similarly find the container volume to be linearly proportional to the absolute temperature T, as shown in Figure 2-4.

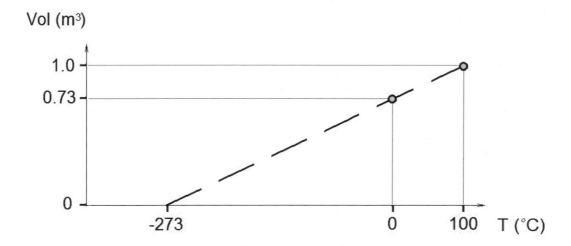

Figure 2-4: Measure changes in volume to determine the absolute temperature scale. By linear extrapolation find that at volume approaches 0 m³, then the absolute temperature, T = -273 °C or 0 K.

It is often required of the engineer to convert a temperature in one unit system to another before all his or her calculations can be completed. Common temperature conversion formulae are

$$T(K) = T(°C) + 273 \tag{2.8}$$

$$T(°R) = T(°F) + 460 \tag{2.9}$$

$$T(°C) = \left(\frac{5}{9}\right)\left[T(°F) - 32\right] \tag{2.10}$$

$$T(°F) = \left(\frac{9}{5}\right)T(°C) + 32 \tag{2.11}$$

Since temperatures on the absolute (K and °R) scale are both zero at the same temperature,

$$T(K) = \left(\frac{5}{9}\right)T(°R) \tag{2.12}$$

$$T(°R) = \left(\frac{9}{5}\right)T(K) \tag{2.13}$$

EXAMPLE 2.1: TEMPERATURE CONVERSIONS

Given:

A thermometer reveals that air in a tank is 72 °F.

Find:

Calculate the temperature of the air in °R, °C, and K.

Solution:

$$T = 72°F + 460 = \boxed{532°R}$$

$$T = \left(\frac{5}{9}\right)[72°F - 32] = \boxed{22.2°C}$$

$$T = 22.2°C + 273 = \boxed{295.2\,K}$$

Note:

$$T = \left(\frac{5}{9}\right)532°R = 295.5\,K$$

Check:

The numerical values and the units appear to be reasonable and address the original question(s).

ENGINEERING IN CODE: TEMPERATURE CONVERSION PROGRAM

The following MATLAB® script may be used for simple conversion between °F, °R, °C, and K:

```
clear all;
T_1 = input('Enter a value of temperature to convert:    ');
F = sprintf('%cF',char(176));
R = sprintf('%cR',char(176));
C = sprintf('%cC',char(176));
choice_1 = menu('Convert from',F,R,C,'K');
choice_2 = menu('Convert to',F,R,C,'K');
S = [F R C 'K '];
if choice_1 == choice_2
    T_2 = T_1;
elseif choice_1 == 1
    if choice_2 == 2
        T_2 = T_1 + 460;
    elseif choice_2 == 3
        T_2 = (5/9)*(T_1 - 32);
    else
        T_2 = (5/9)*(T_1 + 460);
    end
elseif choice_1 == 2
    if choice_2 == 1
        T_2 = T_1 - 460;
    elseif choice_2 == 3
        T_2 = (5/9)*T_1 - 273;
    else
        T_2 = (5/9)*T_1;
    end
elseif choice_1 == 3
    if choice_2 == 1
        T_2 = (9/5)*T_1 + 32;
    elseif choice_2 == 2
        T_2 = (9/5)*(T_1 + 273)
    else
        T_2 = T_1 + 273;
    end
else
    if choice_2 == 1
        T_2 = (9/5)*T_1 - 460;
    elseif choice_2 == 2
        T_2 = (9/5)*T_1;
    else
        T_2 = T_1 - 273;
    end
end
S_1 = S((2*choice_1 - 1):2*choice_1);
S_2 = S((2*choice_2 - 1):2*choice_2);
result = [sprintf('%0.2f ',T_1) S_1 sprintf(' = %0.2f ',T_2) S_2];
disp(result);
```

Example input and resultant output (note MATLAB® interactive pop-up menus not shown):
```
Enter a value of temperature to convert:    45
45.00 °C = 113.00 °F
```

2.6   EQUATION OF STATE

The conclusion from the previous "paper experiments" is that at constant mass, m, and constant container volume, V, the gas density, $\rho = m/V$, is also constant. In that case the absolute pressure, p, is directly proportional to the absolute temperature, T. Further at constant mass, m, and constant pressure, the volume, V, and thus gas the density, ρ, is proportional to the absolute temperature. This results in the Equation of State for an Ideal Gas. The limitation of an "ideal gas" is that it ignores the non-linear relationship between internal energy and temperature. This non-linearity is caused by intermolecular forces that are significant near the boiling temperature, but can be neglected above that critical temperature (which, for example, for oxygen $T_{boil, O2}$ = -118 °C). The ideal gas law provides the following relationship

$$p V = n R_u T \tag{2.14}$$

The molecular weight, M_{gas}, and the number of moles, n_{gas}, of a gas provide its total mass, m_{gas}, as

$$m_{gas} = \left(n M \right)_{gas} \tag{2.15}$$

Furthermore, it is often helpful to use the specific gas constant, R, in place of the universal gas constant, R_u. The universal and specific gas constants are related by the gas' molecular weight as

$$R_{gas} = \frac{R_u}{M_{gas}} \tag{2.16}$$

where

$$R_u = 8314.3\, N \cdot m / kg \cdot K = 1545.4\, ft \cdot lb_f / lbmol \cdot °R$$

Combining Eq. (2.7), Eq. (2.14), Eq. (2.15), and Eq. (2.16), one may derive the equation of state (also known as the "State Equation")

$$p = \rho R T \tag{2.17}$$

As an example, the molecular weight of nitrogen (N_2) is 28 kg/kmol and of for oxygen (O_2) it is 32 kg/kmole. For air, which is composed of 21% O_2 and 79% N_2, its molecular weight is then M_{air} = 0.21 x 32 + 0.79 x 28 = 29 kg/kmol. Aerodynamicists often work with air, so it is handy to know that

$$R_{air} = 287\, N \cdot m / kg \cdot K = 1716\, ft \cdot lb_f / slug \cdot °R$$

It is also worthwhile to note that a simple way of calculating specific gas constants for gases other than air is to multiply the specific gas constant of air by the ratio of the new molecular weight to the molecular weight of air. Evaluating the relationship between the universal and specific gas constants for air and also for some other gas, denoted by the generic subscript gas, one may show that

$$R_{gas} = \left(\frac{M_{air}}{M_{gas}} \right) R_{air} \tag{2.18}$$

TEST YOU UNDERSTANDING: RELATING MOLECULAR WEIGHT AND GAS CONSTANTS

2 grams (2 g) of an unknown compressed gas are tested in a sealed, rigid container with a volume of exactly 1 cubic decimeter. At a measured pressure of 234.7 kPa, the temperature of the gas is found to be 1853 K.

Find the molecular weight of the unknown gas, M_{gas}, in kg per kmol.

Answer: $M_{gas} = 131.3\,kg/kmol \leftarrow$ if it is made up of only one element, the gas is Xenon (Xe)

EXAMPLE 2.2: CALCULATION OF STATE VARIABLES ON A STANDARD DAY

Given:

At sea level (altitude h = 0) air on a "standard day" has a defined temperature of 15.16 °C and pressure p = 101,325 N/m² (Pa or pascal).

Find:

a) What is the air density, ρ?
b) What is the specific volume, $v = 1/\rho$?

Provide answers in both SI and English units.

Solution:

a)

$$SI: \quad \rho = \frac{p}{RT} = \left(\frac{101,325\,N}{m^2}\right)\left(\frac{kg \cdot K}{287\,N \cdot m}\right)\frac{1}{(15.16+273)K} = \boxed{1.225\,kg/m^3}$$

$$Eng: \quad T = \left[15.16\left(\frac{9}{5}\right)+32\right]°F = 59.3°F$$

$$\rho = \frac{p}{RT} = \left(\frac{2116\,lb_f}{ft^2}\right)\left(\frac{slug \cdot °R}{1716\,ft \cdot lb_f}\right)\frac{1}{(59.3+460)°R} = \boxed{0.002376\,slug/ft^3}$$

b)

$$SI: \quad v = \frac{1}{\rho} = \frac{m^3}{1.225\,kg} = \boxed{0.816\,m^3/kg}$$

$$Eng: \quad v = \frac{1}{\rho} = \frac{ft^3}{0.002376\,slug} = \boxed{421\,ft^3/slug}$$

Check:

The numerical values and the units appear to be reasonable and address the original question(s).

ENGINEERING IN SPREADSHEETS: FINDING DENSITY AND SPECIFIC VOLUME

The following Microsoft® Excel® spreadsheet example demonstrates its use in calculating the ambient density and specific volume of a gas after acquiring several user input values. The values are calculated using Eq. (2.17). An excerpt of the spreadsheet is shown in Figure 2-5.

	A	B	C	D	E	F
1		Variable	Symbol	Units	Value	Formula
2	Gas Constants	Universal gas constant	R_u	(N·m/kmol·K)	8314	N/A (fixed value)
3		Gas molecular weight	M_{gas}	(kg/kmol)	29	N/A (user input)
4		Specific gas constant	R_{gas}	(N·m/kg·K)	286.7	=E2/E3
5	Ambient Conditions (knowns)	Pressure	p_{amb}	(atm)	0.96	N/A (user input)
6		Pressure	p_{amb}	(Pa)	97272	=E5*101325
7		Temperature	T_{amb}	(°C)	12	N/A (user input)
8		Temperature	T_{amb}	(K)	285	=E7+273
9	Calculated Values (unknowns)	Density	ρ_{amb}	(kg/m³)	1.1905	=E6/(E4*E8)
10		Specific volume	v_{amb}	(m³/kg)	0.8400	=1/E9

Figure 2-5: Spreadsheet excerpt example for use in determining ambient density and specific volume.

Tips and instructions for the development and use of this spreadsheet:

1) Type in the descriptive names to head each column (in this case, all the names shown in Row 1).
2) Merge the appropriate cells in the first column and enter the descriptive names of the sections.
3) Enter in the names and symbols of the variables in Columns B and C, and their respective units in Column D.
4) Insert and Symbol toolboxes allow the use of Greek characters or symbols (e.g., ρ or °).
5) The Font toolbox allows users to make superscripts or subscripts (e.g., m³ or R_{gas}).
6) To get the "interpunct" (mid dot symbol, e.g., in N·m), users should hold ALT + 0, 1, 8, 3.
7) Make areas of user inputs (in this case with cells filled in green) distinct from areas of calculable output (in this case with cells filled in yellow) to help identify cells of special interest.
8) In the cells where calculations are to be performed, initiate the calculation by the "=" symbol. Example: In cell E6, the value of the cell should be in units of Pa, so the correct way to calculate this would be

$$p_{amb}(Pa) = p_{amb}(atm)\left(\frac{101,325\,Pa}{1\,atm}\right) \qquad \Leftrightarrow \qquad E6 = E5*101325$$

Typing exactly "=E5*101325" in cell E6 therefore causes it to show the value "97272" indicating that a pressure of 0.96 atm is the equivalent of 97,272 Pa.
9) Check to be sure all calculated values make sense.

EXAMPLE 2.3: AIR IN A SEALED FREEZER

Given:

A freezer with known dimensions for its interior size and door size acts as a sealed rigid container whenever its door is closed. Initially (at state 1), the door is allowed to open, filling it with outside air at known conditions. The freezer is then closed to cool the air to its set point refrigeration temperature (at state 2).

$$l_{freezer} = 15\,ft$$

$$w_{freezer} = 18\,ft \qquad w_{door} = 3\,ft$$

$$h_{freezer} = 8\,ft \qquad h_{door} = 7\,ft$$

$$p_{outside} = 1\,atm = 2116\,psfa$$

$$T_{outside} = 95\,°F$$

$$T_{set\,point} = -10\,°F$$

Find:

a) What is the pressure in the freezer at state 2 (in psfa and mmHg)?
b) How many pounds of force are required to open the door at state 2?

Solution:

Many times it helps to picture the problem...

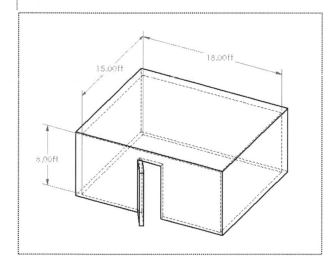

a)

$$p_2 = (\rho R T)_2 = \left(\frac{m}{\forall}\right)_2 R_{air} T_2$$

$$R_{air} = 1716\,ft \cdot lb_f / slug \cdot °R$$

$$T_2 = -10\,°F + 460 = 450\,°R$$

$$\left.\begin{array}{l}"sealed" \to m = const.\\"rigid" \to \forall = const.\end{array}\right\} \to \rho_2 = \rho_1$$

$$\rho_1 = \frac{p_1}{RT_1} = \frac{2,116\,lb_f}{(95+460)\,°R \cdot ft^2}\left(\frac{slug \cdot °R}{1716\,ft \cdot lb_f}\right)$$

$$\rho_1 = 0.002222\,slug / ft^3 = \rho_2$$

$$p_2 = \rho_2 R_{air} T_2 = \left(\frac{0.002222\,slug}{ft^3}\right)\left(\frac{1716\,ft \cdot lb_f}{slug \cdot °R}\right)450\,°R = \boxed{1,716\,psfa}$$

$$p_2 = 1,716\,psfa\left(\frac{760\,mmHg}{2,116\,psfa}\right) = \boxed{616\,mmHg}$$

b)

$$F_{door} = \sum pA_{door} = (p_{outside} - p_{inside})A_{door} = (p_1 - p_2)w_{door}h_{door}$$

$$F_{door} = \frac{(2,116 - 1,716)lb_f}{ft^2}(3\,ft)(7\,ft) = \boxed{8,400\,lb_f}$$

Check:

The numerical values and the units appear to be reasonable and address the original question(s). It is interesting to note the amount of force required to open the door is quite large, but for anyone who has ever used such freezers they are likely aware of the fact that the freezers are not in fact air tight and the door handle is specially designed to provide the leverage force necessary to "break the seal" on the door so that the inrushing air can quickly equalize the force on both sides of the entrance door.

2.7   BUOYANCY IN FLUIDS

The careful reader may have already started to suspect that varying the values of pressure, temperature, and especially density may have a significant impact on the relative weight of a packet of gas. Furthermore, manipulation of this fact can lead to several useful tools for the aerospace engineer. Consider a typical hot air balloon like that of the Montgolfier brothers: the deliberate manipulation of the temperature within the flexible balloon caused the air inside it to exhibit significantly lower density than the ambient air on the outside of the balloon. This is because the volume of the balloon is essentially constant, and the pressure outside the balloon must be equivalent to the air outside the balloon (otherwise, it would either expand or collapse since it is flexible), but the temperature inside the balloon is greater than the temperature outside it. Thus, in accordance with Eq. (2.17), the density inside the balloon must be less than the density outside the balloon. As a result of the difference in densities, the lighter gas must rise. This effect is known as buoyancy.

Almost everyone is familiar with the phenomenon of buoyancy. Archimedes of Syracuse, perhaps the greatest mathematician of all time, described this behavior in his treatise *On Floating Bodies*, sometime around 250 BC. In the first of this two-book volume, he established what is known today as Archimedes' principle:

> *Any body placed wholly or partially into a fluid will experience an expelling force equal to the weight of the fluid the body has displaced.*

What Archimedes' principle dictates, then, is that there is always some "attempt" made by a fluid to push objects out of itself with a measurable force that is exactly caused by the fact that fluid has to be raised to insert the object into it. Mathematically, the buoyancy force, F_B, may be expressed as

$$F_B = W_{\substack{fluid \\ displaced}}$$ (2.19)

What is most interesting is that for gases (which always take up the shape of the container in which they are held), Eq. (2.19) may be further refined as

$$F_B = g\left(\rho \forall\right)_{\substack{fluid \\ displaced}}$$ (2.20)

EXAMPLE 2.4: DETERMINIATION OF FLUID DENSITY USING A KNOWN WEIGHT

Given:

A piece of 6061-T6 aluminum with known physical and material properties is suspended above a pool of unknown inert liquid by a cable equipped with a load cell to indicate the tension on the cable. At time 1, the tension, T, caused entirely by the weight of the aluminum block is recorded while it is held out of the liquid. The block is then completely immersed in the liquid and the tension on the cable is recorded again at time 2. The following is known:

$$g = 9.80\,m/sec^2 \qquad \rho_{Al} = 2.70\,kg/m^3 \qquad T_1 = 42.3\,kN$$
$$\Psi_{Al} = 1.60\,m^3 \qquad T_2 = 21.8\,kN$$

Find:

a) What is the density of the unknown liquid (in kg/m^3)?
b) What is the specific gravity of the liquid?

Solution:

Assume the cable supporting the block does not significantly displace any liquid. Using Eq. (2.20),

a)

$$T_1 = W_{Al} \qquad T_2 = W_{Al} - F_B \quad \Rightarrow \quad F_B = T_1 - T_2$$

$$\rho_{fluid \atop displaced} = F_B / \Psi_{Al}\,g = (T_1 - T_2) / \Psi_{Al}\,g$$

$$\rho_{fluid \atop displaced} = (42{,}300 - 21{,}800)\,N \left(\frac{1}{1.60\,m^3} \right) \left(\frac{sec^2}{9.80\,m} \right) \left(\frac{kg \cdot m}{N \cdot sec^2} \right) = \boxed{1307\,kg/m^3}$$

b)

$$S.G._{fluid \atop displaced} = \rho_{fluid \atop displaced} / \rho_{H_2O} = 1307/1000 = \boxed{1.31}$$

Check:

The answers' magnitudes are reasonable, and the units are in agreement with the posed questions.

TEST YOUR UNDERSTANDING: BOUYANCY OF A BOWLING BALL

A bowling ball weighing 15 lb$_f$ was bowled right off the end of a fishing pier at an ocean side resort. A local fisherman snags the ball with his hook, but only has 5 lb$_f$ test fishing line on his fishing rod (i.e., the line will break if it pulls more than 5 lb$_f$).

If the density of the seawater is 1.99 slug/ft³ and the volume of the bowling ball is 321 in³, will the fisherman be able to bring the ball to the surface so that he can grab it with his net?

Answer: Yes, it turns out that the ball will only weigh about 3.1 lb$_f$ while immersed in the seawater.

2.8   GAS MOVING AT A BULK VELOCITY

It is important to understand that for the previous discussions, the gas was limited to a stationary or non-moving condition inside a container. When a gas exhibits a velocity (either localized motion or bulk motion) state variables will be influenced. Gas bulk velocity, V, is a vector with both magnitude and direction. Its directional bulk KE, $V^2/2$, per unit mass can be superimposed on the random molecular motion kinetic energy, $c^2/2$. Then the gas total kinetic energy is the sum of the random motion and directional velocity (i.e., KE = $V^2/2 + c^2/2$). When air flows over an aerodynamic body its frontal area represents an obstruction to the flow that forces the air to take a detour around the body. To maintain a constant rate of air mass flow rate, the air must speed up in the space remaining adjacent to the body. This behavior is evident in a host of everyday events. In a river containing a large boulder, it is easy to see that the water speeds up while flowing past the obstruction presented by the boulder. Another observation is that where the water flows faster next to the boulder, the water surface level drops, indicating a reduction in what is known as hydrostatic pressure (this topic will be discussed in later chapters). Likewise in a gas, because the total energy of the gas is conserved, when bulk kinetic energy increases then internal random motion kinetic energy must drop, which means the temperature and pressure also drop. This fact establishes a critical relationship between pressure, temperature, and velocity that must be ingrained in every aerospace engineer's basic understanding of gases in motion: as bulk velocity increases, temperature and pressure decrease; as bulk velocity decreases, temperature and pressure increase. Simply stated,

$$V \uparrow \;\Rightarrow\; p \downarrow, T \downarrow$$
$$V \downarrow \;\Rightarrow\; p \uparrow, T \uparrow$$

This special relationship between the local static state variables and the local velocity will show up in many different equations aerospace engineers commonly use, and are especially obvious in the Bernoulli and energy conservation equations that are covered later in this book.

When airflow stagnates (i.e., comes to a halt) on the forward-most portion of an airplane component, such as the leading edge on a wing, the local velocity, V, there goes to zero. There the lost directional KE, $V^2/2$, is converted over to random motion K.E., $c^2/2$, and therefore the gas temperature, T, rises to the "stagnation temperature" called T_0. With this rise in temperature, the speed, c, and thus momentum of the molecules prior to each collision increases, which causes the pressure to rise to the "stagnation pressure" p_0. From the inverse relationship between velocity and static state variables, it is possible to conclude that the greatest magnitude of possible static temperature and pressure must occur when the local velocity is zero. This idea establishes yet another critical relationship about temperatures and pressures in a flow:

$$V = 0 \;\Rightarrow\; \begin{cases} p = p_0 = p_{max} \\ T = T_0 = T_{max} \end{cases}$$

EXAMPLE 2.5: MEAN VELOCITY OF HELIUM PARTICLES IN A CONTAINER

Given:

The lightest noble gas, Helium (He), exists in a monatomic molecular structure. In this case, He gas particles are held in a sealed container. The specific internal kinetic energy, *ke*, of the He is readily calculated by the mean velocity, *c*, of the molecules and is proportional to the absolute temperature of the gas by a factor of 1.5 times the specific gas constant for He such that

$$\frac{c^2}{2} = \frac{3}{2} R_{He} T \qquad\qquad M_{He} = 4.003\, kg/kmol \qquad\qquad T = 59°F$$

Find:

Calculate the mean velocity (in *ft/sec*) of the helium molecules at the given temperature.

Solution:

$$c^2 = 3R_{He}T \qquad\qquad c = \sqrt{3R_{He}T}$$

$$R_{He} = \left(\frac{M_{air}}{M_{He}}\right) R_{air} = \left(\frac{29}{4}\right) 1{,}716\, ft \cdot lb_f/slug \cdot °R = 12{,}441\, ft \cdot lb_f/slug \cdot °R$$

$$c = \sqrt{3\left(\frac{12{,}441\ ft \cdot lb_f}{slug \cdot °R}\right)(59+460)°R \left(\frac{slug \cdot ft}{lb_f \cdot sec^2}\right)} = \boxed{4{,}359\ ft/sec}$$

Check:

The numerical values and the units appear to be reasonable and address the original question(s).

TEST YOUR UNDERSTANDING: AIR INSIDE A SPINNING VORTEX

Air in a tightly-bound vortex spins in solid body rotation (i.e., the fluid acts as if it is a rigid body where the velocity varies proportionally with radius as measured from the center of the vortex).

Where can one find the stagnation pressure and temperature of the flow, and where is the minimum pressure and temperature in the flow? Explain why this is so.

Answer: Stagnation values are at the vortex core where the radius, thus velocity, is zero; minimum values are at the outer vortex boundary where the radius, thus velocity, is at its maximum value.

In cases where surface friction is negligible, the integration of pressure over aerodynamic surfaces provides the aerodynamic forces of interest. For a wing to produce lift, it must accelerate the air over its upper surface to a higher velocity than over its lower surface. Therefore, wings are cambered, giving more curvature on top of the wing than on the bottom. The result is a higher upper surface velocity, thus lower pressure on the upper surface than on the lower surface. This difference literally lifts the wing up.

A gas inside a container exerts only a normal pressure force to the walls. However in a flowing gas, molecules lose directional K.E. near the surface due to skin friction, altering the direction of the collision reflections. In the process, the gas molecules transfer some of their momentum to the wall. This produces drag in the downstream direction. This drag force is usually much smaller than the lift caused by pressure. Its magnitude per unit area is given by the shear stress, τ, which has the same units as pressure but acts tangentially to the surface, while pressure acts perpendicular to the surface.

An airfoil is a two-dimensional (i.e., infinite-span) wing. When mounted in between the walls of a wind tunnel, it can be used to study the flow of air over the wing. When smoke is injected from small tubes upstream, the trajectory of the air particles shows up as "streamlines", as shown in Figure 2-6. Shown is a highly efficient airfoil shape as used on the first human powered airplane, the Gossamer Condor, designed by Paul MacCready and Peter Lissaman of AeroVironment, Inc. Another famous airfoil designer, Robert Liebeck, is a well-known aerodynamicist, consultant to the American Cup racer designers, and the originator of the Liebeck airfoil series. The Liebeck high lift to drag ratio airfoil shape is used as the tail-wing on most Indy 500-race cars to provide downward lift for increased traction.

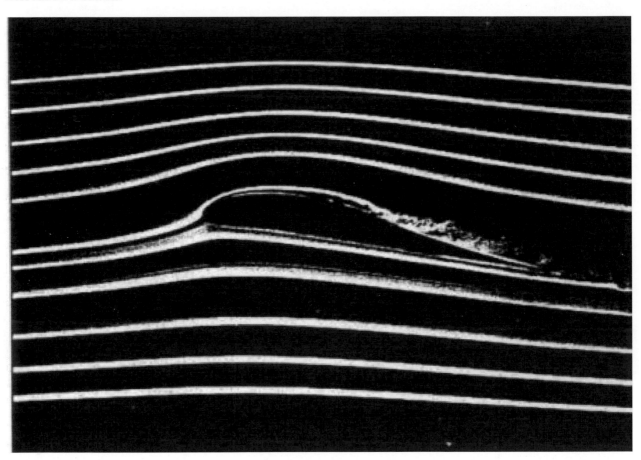

Figure 2-6: Smoke photograph of the low-speed flow over a Lissaman 7769 airfoil at 10° angle of attack.

In flight, the only aerodynamic forces which can act on an airfoil are the surface integrals of the pressure, p, and the wall shear stress, τ_w. Pressure is due to changes in velocity and acts perpendicular to the surface while shear stress is due to friction acting in the downstream direction tangential to the surface (Figure 2-7).

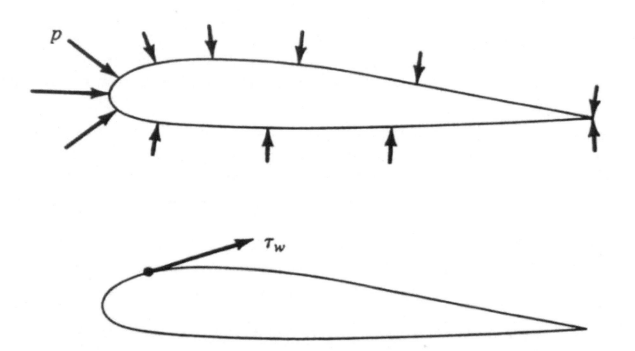

Figure 2-7: Pressure, p, is the perpendicular force component per unit area acting on the surface and sheer stress at the surface (or "wall"), τ_w, is the force component per unit area tangential to the surface.

2.9   SUMMARY

UNITS

In engineering, it is always best to express units in fundamental, absolute terms (N, kg, m, sec, K or lb$_f$, slug, sec, °R). Keep in mind that Newton's second law provides the key to converting units between m-L-t and F-L-t systems.

IDEAL GAS AND THE EQUATION OF STATE

An ideal gas is a gas that, regardless of the situation, obeys the relationship

$$p V\!\!\!\!/ = nR_u T$$

One can relate the universal and specific gas constants as

$$R_{gas} = \frac{R_u}{M_{gas}}$$

The standard form of the equation of state can thus be expressed as

$$p = \rho RT$$

Pressure – this is a way to describe the collective effect of the collisions normal to a surface of all the innumerable molecules of gas next to it.

Density – this is the way to describe the collective amount of gas matter within a specified volume.

Specific gas constant – this is the way to describe the energy content of the gas one is working with at a known temperature per unit mass.

Temperature – this is the way to describe the energy content associated with the mean random molecular kinetic energy of the gas one is working with.

MANOMETERS AND ABSOLUTE PRESSURES

For any column of fluid in a manometer,

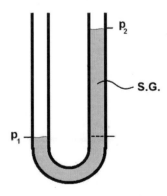

$$p_1 + \rho g h_1 = p_2 + \rho g h_2$$

Terms may be rearranged such that

$$p_1 = p_2 + \rho g (h_2 - h_1) = p_2 + \rho_{fluid} g \Delta h_{fluid}$$

Noting that specific gravity, S.G., defines a fluid's density as a ratio with respect to the density of water,

$$S.G. = \frac{\rho_{fluid}}{\rho_{H_2O}} \qquad \Rightarrow \qquad \rho_{fluid} = S.G. \left(\rho_{H_2O} \right)$$

The typical expression for manometers may be expressed as

$$p_1 = p_2 + S.G. \left(\rho_{H_2O} \right) g \Delta h_{fluid}$$

The density of water is easy to remember:

$$\rho_{H_2O} = 1000 \, kg/m^3$$

Remember that most of the time engineers work in absolute pressures. There are sometimes good reasons to use for gauge pressures, however, which can be related to absolute pressures in the following way:

$$p_{abs} = p_g + p_{atm} = const.$$

BUOYANCY PROBLEMS

The driving force of buoyancy is the fluid displaced by the object. When an object is inserted into a fluid, the fluid is raised slightly, and the difference between the weights of that fluid and one taking its place is exactly the force trying to drive the object back out of the fluid, or

$$F_B = W_{fluid \, displaced} - W_{fluid \, filling}$$

For gases, this may also be written

$$F_B = \left(\rho_{fluid \, displaced} - \rho_{fluid \, filling} \right) Vg$$

MOVING GASES

The most important concept to remember is that when a gas moves in bulk, it exhibits a change in static pressure and temperature. As velocity increases, static pressure and temperature decrease. As velocity decreases, static pressure and temperature increase.

$$V \uparrow \Rightarrow p \downarrow, T \downarrow$$
$$V \downarrow \Rightarrow p \uparrow, T \uparrow$$

Pressure and temperature are at their maximum values when the fluid is stagnant.

CHAPTER 2 PROBLEMS

2.1) Measuring the drop in pressure with a manometer as a function of temperature within a sealed container filled with an ideal gas provides a technique to show that the magnitude of "zero absolute temperature" or 0 K corresponds to -273 °C. If this experiment is conducted, starting at 1 atm and 67 °C, instead of in boiling water, what will the pressure be, in mmHg, when cooled to 0 °C?

2.2) An air tight freezer measures 4 m x 5 m x 2.5 m high. With the door open, it fills with 22 °C air at 1 atm pressure.

    a. Calculate the density of this air in kg/m³
    b. After closing the door it is cooled down to 7 °C. How low will the pressure in the freezer be in units of Pa and mmHg?
    c. How many newtons of force will be needed to open the 1 m x 2 m door?

2.3) A space shuttle moves through the atmosphere on its return from space. At point 1 on its wing, the air temperature and pressure are measured to be $T_1 = 300°C$ and $p_1 = 10,000$ Pa while at point 2 they are $T_2 = 700°C$ and $p_2 = 575,000$ Pa. Calculate the air density in kg/m³ at both points 1 and 2.

2.4) Consider a 1.5 cubic feet scuba tank filled with air at 2250 psia at $T = 70$ °F.

    a. Calculate the mass of air, m, in the tank in slugs and in lb_m.
    b. Calculate the standard cubic feet (SCF) of air inside the tank (Note: "standard cubic feet" is the volume the gas would take up in a standard atmosphere. SCF = mass divided by sea level air density, ρ_{sl}).

2.5) Assume the shape of a helium balloon to be a sphere of radius, r, equal to 9 m, at standard sea level conditions. The volume, V, of a sphere is expressed as $V = (4\pi/3)r^3$. Note the molecular weight of He is $M_{He} = 4$ kg/kmol and for air, $M_{air} = 29$ kg/kmol.

    a. Calculate the density of the helium, ρ_{He}, inside balloon and the weight of the balloon, W, assuming the material of the balloon is negligible.
    b. Calculate the balloon's buoyant force, F_B.
    c. What is the name of the principle by which the helium balloon derives its lift force?

2.6) If gas molecules have a mean random motion velocity, c, and their internal kinetic energy equals $0.5c^2 = 1.5RT = 1.5(R_u/M)T$, what is mean velocity c in (ft/s) for the helium molecules inside a balloon at standard sea level temperature of 59°F.

2.7) On a standard day at sea level an airplane tire is pumped up to a tire pressure gauge reading of 40 psig. After landing on a mountaintop where the ambient air pressure is 8 psia, the pilot checks the tires again with the gauge. What will the pressure gauge indicate?

2.8) On a standard day a mercury manometer is used to measure the pressure in a gas pipeline located at sea level. The manometer leg, which is open to the atmosphere, has its mercury level 1.5 m higher than the level of mercury in the manometer leg connected to the gas pipeline. Determine the pressure, p, in the pipeline in both Pa and in meters of water.

CHAPTER 3.   ORBITAL MECHANICS AND THE ATMOSPHERE

3.1   INTRODUCTION

Celestial mechanics is the study of the motion of the extremely large objects commonly referred to as celestial objects (e.g., Earth, Sun, bodies outside the Earth's solar system, and so on). Orbital mechanics further refines the discussion of object motion to include manmade objects such as rockets or satellites. While it may not be immediately obvious, some of the underlying fundamental principles related to the study of massive objects far away from Earth also have a critical impact on the physical state of tiny objects in close proximity to the Earth's surface. How can celestial and terrestrial physics be described using the same basic physical laws? It turns out that gravitational acceleration is as directly responsible for the motion of Earth with respect to the Sun as it is responsible for the stratified layers of life-sustaining gases held close to the Earth's surface that is generically referred to as its atmosphere. This chapter will discuss the impact of gravitational acceleration both inside and outside the Earth's atmosphere.

3.2   OUTSIDE THE ATMOSPHERE

Earth's radius, r_e, is quite large and greater at the equator than at the poles due to the centripetal force caused by revolving about its axis. For calculation purposes, assume the Earth to be spherical with a constant radius equal to the radius at 45° latitude, where r_e = 6,357 km. The atmosphere is approximately 160 km, or approximately 100 miles, thick. Thus the percentage of thickness of the atmosphere with respect to the radius of the Earth is about 2.5%. At the equator, where one nautical mile is equal to 1 minute longitude (1852 m), the radius of the Earth is larger and is found as (1852 m per minute x 360 degrees per circumference x 60 minutes per degree)/(2π) = 6,367 km.

To launch a satellite, a carrier rocket takes off vertically from Earth. It first passes through the lower atmosphere called the troposphere, which reaches up to an altitude of 11 km. All airline travel takes place within this region. The rocket then it passes through the stratosphere which goes from an altitude of 11 km to an altitude of 25 km. Only specialized planes like the U-2 and SR-71 can operate within this region of the Earth's atmosphere. Next it enters the mesosphere from 25 km to 47 km. In this region, only rockets can operate because the atmospheric oxygen is too sparse for use as a fuel oxidizer in an internal combustion engine. Rockets perform fine here because they carry the required oxidizer on board and therefore can produce propulsive thrust even in outer space. The X-15 rocket plane made it to the upper atmosphere at 95 km altitude in the 1960's, and in the early 2000's, SpaceShipOne made it to 112 km. However, to actually orbit the earth at minimum drag, one must climb higher, to an altitude of at least 200 km. Even there in outer space, occasionally a burst of thrust will be required to maintain altitude. There satellites still experience some drag from collision with freely moving molecules, collectively called inter-stellar gas. Drag is the net force opposing the direction of motion caused by the momentum exchange associated with a change in direction and velocity of a gas molecule when it collides with the satellite surface. This drag is small and therefore often ignored in the analysis of the effect of gravitational forces on the motion of satellites and spaceships. When one travels dozens of Earth radii away, then the Sun's gravitational force begins to overpower that of the Earth. The reason for this is that the mass of the Sun is 332,488 times greater than that of Earth, and controls the greatest portion of interplanetary flight. The orbital radius of the planet Earth about the Sun is 149.5 x 10^6 km.

Sir Isaac Newton first published his ideas on the attractive force between two bodies in 1687. Based upon empirical observations, Newton proposed what is now known as the law of universal gravitation which states that for any two bodies in the universe, the attractive force between them is directly proportional to the product of the two masses and inversely proportional to the square of the distance between the two masses. This force acts between all bodies, but becomes significant particularly when the two objects in question are quite massive or are very close to one another. Objects such as planets and stars are quite massive, and thus tend to be the primary subjects of study when discussing sources of gravitational attraction. These "celestial bodies" are fortunately approximately spherical and can be treated roughly as point masses (i.e., their entire mass can be thought of as being located at a single point, namely the sphere's center which is also its center of mass). For the case of point masses, Newton's law of universal gravitation may be expressed

$$F = G\frac{m_1 m_2}{r^2} \qquad (3.1)$$

where m_1 and m_2 are the values of the two masses, r is the distance between them, and

$$G = 6.674 \times 10^{-11}\ N \cdot m^2/kg^2$$

In the event that one of the two masses is much larger than the other (i.e., $m_1 >> m_2$), then one may express the gravitational force as

$$F = \frac{\mu m}{r^2} \qquad (3.2)$$

where m is the mass of the small object and $\mu \approx Gm_1$. Connecting this idea with Newton's second law, it is possible to express the acceleration on a relatively small object in the presence of a massive object as

$$g = \frac{\mu}{r^2} \qquad (3.3)$$

Consider the kinetic energy necessary to make sure that an object is able to effectively exit the gravitational field of a nearby massive object. Since both the potential energy associated with the field and the kinetic energy of the small object are directly proportional to its mass, this problem reduces to a question of critical minimum speed. This value is known as the escape velocity. Consider the initial and final states of energy for the object, where the initial energy per unit mass is

$$\frac{E_i}{m} = e_i = \left(-gh_{a,0} + \frac{V_0^2}{2}\right) = \left(-\frac{\mu h_{a,0}}{r^2} + \frac{V_0^2}{2}\right) = \left(-\frac{mG}{r} + \frac{V_0^2}{2}\right)$$

The negative sign leading the potential energy term indicates the reducing strength of the field as the distance between the two objects is increased. The final energy per unit mass is assumed to be zero since the object must eventually come to rest at an infinite distance from the gravitating body, thus

$$\frac{E_f}{m} = e_f = \left(-gh_{a,f} + \frac{V_f^2}{2}\right) = \left(-\frac{mG}{\infty} + \frac{0^2}{2}\right) = 0$$

Energy must be conserved, thus by equating the initial and final conditions it is easy to find the minimum speed required for the object to leave the gravitational influence of the large body as

$$V_e = \sqrt{2mG/r} \tag{3.4}$$

EXAMPLE 3.1: ESTIMATE THE MASS OF THE EARTH

Given:

A cube of steel (ρ_{steel} = 8,000 kg/m^3) with a side length of 10 cm has a known weight of 78.4 N on Earth at sea level where the Earth's radius is approximately 6,371 km.

Find:

Estimate the mass of the Earth using the law of universal gravitation.

Solution:

$$F = Gm_{Earth}m_{steel}/r_e^2 = m_{steel}g_0 = W_{steel}$$

$$m_{Earth} \approx \frac{r_e^2 W_{steel}}{m_{steel}G} = \left(\frac{6,357,000^2\,m^2}{8\,kg}\right)\left(\frac{78.4\,N}{}\right)\left(\frac{kg^2}{6.674\times10^{-11}\,N\cdot m^2}\right) = \boxed{5.93\times10^{24}\,kg}$$

Check:

According to NASA, the mass of the Earth is 5.9726 x 10²⁴ kg.

EXAMPLE 3.2: CALCULATE THE ESCAPE VELOCITY OF THE EARTH'S MOON

Given:

The Earth's moon has a mass of 7.342 x 10²² kg and a mean radius of 1,737 km.

Find:

Calculate the escape velocity required to leave the moon's gravitational field (neglect Earth's gravitational pull on the escaping object).

Solution:

$$V_{e,moon} = \sqrt{\frac{2m_{moon}G}{r_{moon}}} = \left[2\left(\frac{7.342\times10^{22}\,kg}{}\right)\left(\frac{6.674\times10^{-11}\,N\cdot m^2}{kg^2}\right)\left(\frac{1}{1,737,000\,m}\right)\right]^{\frac{1}{2}} = \boxed{2,375\,m/\sec}$$

Check:

According to NASA, the escape velocity of the Earth's moon is 2.38 km/sec.

Some interesting physical values for celestial bodies in the Earth's solar system are presented in Table 3-1.

Table 3-1: Physical values for celestial objects in Earth's solar system

Object	Mean Radius (km)	Relative Mass†	Surface Gravity (m/sec²)	Escape Velocity (m/sec)
Sun	696,342	332,900	274.090	617,500
Mercury	2,440	0.055	3.700	4,217
Venus	6,051	0.815	8.870	10,360
Earth	6,357	1.000	9.805	11,180
Moon	1,737	0.012	1.622	2,375
Mars	3,396	0.107	3.711	5,032
Jupiter	71,400	317.800	24.790	59,568
Saturn	60,000	95.110	10.440	35,550
Uranus	25,400	14.520	8.690	21,350
Neptune	24,300	17.140	11.150	23,710
Pluto	1,500	0.002	0.658	1,229

† Mass of the Earth is 5.9726 x 10²⁴ kg

The acceleration of gravity on the surface of the Earth at sea level is

$$g_0 = 9.81\,m/\sec^2 = 32.2\ ft/\sec^2$$

In circular orbit, the absolute altitude, h_a, is defined as the radial distance from the center of the Earth. By Newton's law, the local acceleration of gravity, g, is inversely proportional to the square of the radial distance, h_a, from its center. It is possible to write Eq. (3.3) in a form which relates the local gravitational acceleration to that of the Earth's surface such that

$$g = g_0 \left(\frac{r_e}{h_a} \right)^2 = g_0 \left(\frac{r_e}{r_e + h_G} \right)^2 \tag{3.5}$$

Note that Eq. (3.5) uses another altitude called the geometric altitude, h_G. This is defined as the altitude above mean sea level and is calculated as

$$h_G = h_a - r_e \tag{3.6}$$

One can imagine that geometric altitude, h_G, could be measured anywhere inside the atmosphere by simply dropping a tape measure down to sea level, thus, for example, at sea level $h_G = 0$ km.

To minimize drag, satellites orbit the Earth at a geometric altitude of at least 200 km. This is called Low Earth Orbit (LEO). The centripetal radial force acting on a satellite of mass, m, is given by

$$F = \frac{mV_{orbit}^2}{h_a} \tag{3.7}$$

For equilibrium, this force must be opposed by the weight of the satellite in orbit such that $W = mg$. By equating these two forces it is possible to find the orbital velocity

$$V_{orbit} = \sqrt{gh_a} \qquad (3.8)$$

Combining the known distance from the Earth's center with the calculated orbital velocity, and assuming a circular orbital path, it is possible to calculate the time it takes for an object to complete a single orbit around the Earth. This is known as the object's period, and is calculated as

$$t_{orbit} = \frac{2\pi h_a}{V_{orbit}} \qquad (3.9)$$

EXAMPLE 3.3: CALCULATE THE PERIOD OF A SATELLITE IN ORBIT

Given:

A satellite follows a circular orbit at a geometric altitude of 200 km.

Find:

Calculate the period of the satellite.

Solution:

$$t_{orbit} = \frac{2\pi h_a}{V_{orbit}} \qquad \rightarrow \qquad \text{need } h_a \text{ and } V_{orbit}$$

$$h_a = r_e + h_G = (6,357 + 200)\,km = 6,557\,km$$

$$V_{orbit} = \sqrt{gh_a} \qquad \rightarrow \qquad \text{need } g$$

$$g = g_0 \left(\frac{r_e}{h_a}\right)^2 = 9.81\,m/sec^2 \left(\frac{6,357}{6,557}\right)^2 = 9.22\,m/sec^2$$

$$V_{orbit} = \left[\left(\frac{9.22\,m}{sec^2}\right)\left(\frac{6,557,000\,m}{}\right)\right]^{\frac{1}{2}} = 7,775\,m/sec$$

$$t_{orbit} = \left(\frac{2\pi}{}\right)\left(6,557,000\,m\right)\left(\frac{sec}{7,775\,m}\right) = \boxed{5,298\,sec = 88.3\,minutes}$$

Check:

The numerical values and the units appear to be reasonable and address the original question(s).

The motion induced by gravitational attraction between two bodies follows a predictable path. In the early 1600's, Johannes Kepler was perhaps the first scientific investigator to describe many of the fundamental laws of optics and apply his research to planetary motion. Kepler noted that like light through a lens, planetary motion tends to exhibit a point of focus. Using astronomical observations gathered by other famous thinkers such as Tycho Brahe, Kepler was able to figure out that all planets move according to an elliptical orbit and established Kepler's laws of planetary motion:

1. A planet's orbit follows an elliptical path with the Sun at one of the two foci.

2. For a given time interval, a line between the Sun and a select planet sweep out a specific area, regardless of the planet's initial position.
3. The cube of the semi major axis of the planet's path is directly proportional to the square of the time required for the planet to complete one orbit.

In fact, it is now known that the motions of all satellites follow paths which fall into one of four shapes from a family of curves known as conic sections. A conic section is a curve formed by the intersection of an axisymmetric mirrored cone and a plane (Figure 3-1). Noting that the curve comes from the intersection of a three dimensional surface and a plane, the curve is necessarily two dimensional in nature and thus can be mathematically described in general terms as

$$Ax^2 + Bxy + Cy^2 + Dx + Ey + F = 0 \qquad (3.10)$$

Although it is possible to generate a single line, a pair of intersecting lines, and even a single point by intersecting a mirrored cone with a plane as shown in Figure 3-1, the conic sections most relevant to orbital motion are the circle, ellipse, parabola, and hyperbola. A circle is defined as the locus of all points whose distance to a select center point is constant (Figure 3-2). An ellipse is defined as the locus of all points in which the sum of the distance to each of two foci is constant (Figure 3-3). A parabola is defined as the locus of all points for which the distance to the focus is equivalent to the distance to a line known as the directrix (Figure 3-4). Finally, a hyperbola is defined as the locus of all points for which the difference between the distances from each of two foci is constant (Figure 3-5).

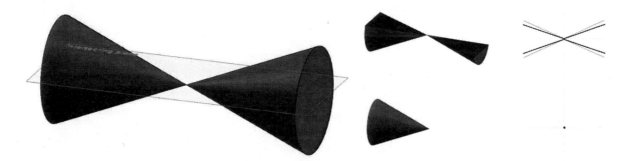

Figure 3-1: A mirrored pair of cones may be intersected by a simple plane to produce a conic section. Single lines, pairs of intersecting lines, points, circles, parabolas, ellipses, hyperbolas, and even parallel lines are all possible conic sections.

One way to think about the four primary types of satellite paths is by considering how close the curve is to being circular. This measure, which is commonly known as eccentricity, e, is zero for a perfectly circular curve, between zero and one for an elliptical curve, exactly one for a parabolic curve, and greater than one for hyperbolic curves. Two key parameters that help determine the eccentricity of conic sections are the semi-major axis, a, and the semi-minor axis, b. The conic sections along with the respective equations for their 2D curves and eccentricities are provided in Figure 3-2 through Figure 3-5. Note that one may easily relate the 2D equations to Eq. (3.10).

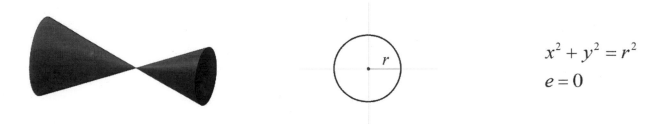

$$x^2 + y^2 = r^2$$
$$e = 0$$

Figure 3-2: A plane normal to the axis of revolution intersecting the cones produces a circular conic section.

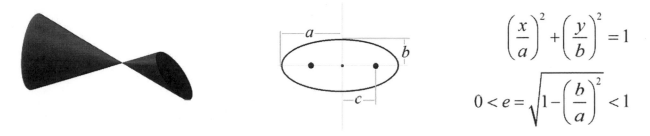

$$\left(\frac{x}{a}\right)^2 + \left(\frac{y}{b}\right)^2 = 1$$
$$0 < e = \sqrt{1 - \left(\frac{b}{a}\right)^2} < 1$$

Figure 3-3: A plane that intersects only one cones produces an elliptical conic section.

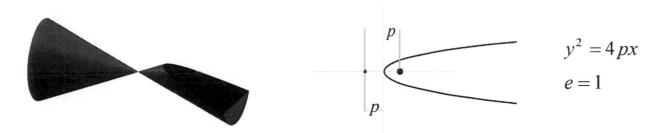

$$y^2 = 4px$$
$$e = 1$$

Figure 3-4: A plane that intersects only one cone at an angle identical to the half angle of the cone produces a parabolic conic section.

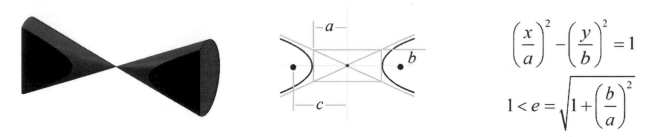

$$\left(\frac{x}{a}\right)^2 - \left(\frac{y}{b}\right)^2 = 1$$
$$1 < e = \sqrt{1 + \left(\frac{b}{a}\right)^2}$$

Figure 3-5: A plane that intersects both cones but does not pass through the points of the cones produces a hyperbolic conic section.

Consider the solar system and three special celestial bodies that operate within it. First, consider the Sun, whose mass is sufficient to maintain the structure of the solar system itself. Next, consider "the green planet", Earth. Finally, consider Halley's Comet. These three objects are presented in Figure 3-6. The time it takes for Earth to complete one orbit, or Earth's period, is 365.25 Earth days. This amount of time is also known as a Julian Year. Earth's eccentricity is 0.0167, indicating its closely

43

circular orbital path. On the other hand, Halley's Comet has a period of 75.3 Julian years and has an eccentricity of 0.967, indicating its highly elliptical orbital path.

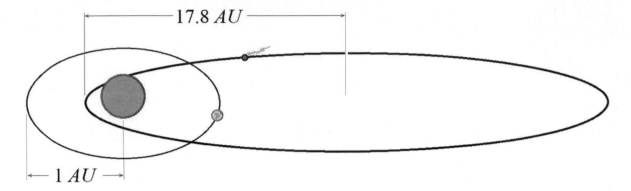

Figure 3-6: Two celestial bodies in orbit about the Sun in our solar system: Earth's orbit is nearly circular while Halley's Comet is highly elliptical.

To help define great lengths such as those that might describe the incredible distances between celestial objects, astronomers, scientists and engineers use what is known as Astronomical Units, or *AU*. This unit of measure is roughly the mean distance from the Earth to the Sun, and is defined as

$$1\,AU = 149{,}597{,}870{,}700\,m \approx 93 \times 10^6\,mi$$

EXAMPLE 3.4: ESTIMATE THE LENGTH OF HALLEY'S COMET'S SEMI-MINOR AXIS

Given:

The length of Halley's Comet's semi-major axis is known to be 17.8 AU, and its eccentricity is known to be 0.967.

Find:

Determine the approximate length of the comet's semi-minor axis.

Solution:

From Figure 3-3, $e = \sqrt{1 - \left(\dfrac{b}{a}\right)^2}$ or $b = a\sqrt{\left(1 - e^2\right)}$

$b = \left(17.8\sqrt{1 - 0.967^2}\right) AU$

$b = \boxed{4.54\,AU}$

Check:

According to NASA, the length of Halley's Comet's semi-minor axis is 4.8 AU.

TEST YOUR UNDERSTANDING: USING KEPLER'S LAWS TO RELATE TWO ORBITS

Earth is known to have a semi-major axis of 149.60×10^6 km and an orbital period of 365.25 Earth days. It is also known that for a full Earth year (i.e., 365.25 Earth days), Saturn moves through exactly 3.3948% of its own orbital period around the Sun.

Using Kepler's laws of planetary motion, determine the length of Saturn's semi-major axis in km.

Answer: Saturn's semi-major axis is approximately 1427×10^6 km

It is worthwhile to consider practical methods of getting from the Earth's surface to a possible point far enough away from the Earth to establish an orbital path like that of a typical commercial or scientific satellite. To minimize the required rocket thrust to reach orbital velocity one can take advantage of the Earth's surface rotational velocity to the East. The Earth itself orbits the Sun in a year and rotates about its own axis every 24 hours. The circumference of our Earth at the equator is calculated from $2\pi r_e = 39{,}942$ km. To make a complete turn in 24 hours requires a surface rotational velocity, $V_{surface} = 39{,}942$ km per 24 hours = 1,664 km/hr = 0.46 km/second moving in an eastward direction. Therefore space agencies launch satellites in the eastern direction to capitalize on the already-high relative motion of the Earth's surface. The European Space Agency launches just north of the equator from the Guiana Space Center in Kourou, French Guiana. The U.S. uses NASA's Cape Canaveral in Florida. Many space agencies (including NASA) also launch missions from the Baikonur Cosmodrome in Kazakhstan. Ideally, facilities overlook large bodies of water to the east which adds the advantage of having debris from launch failures fall onto uninhabited areas.

Atmospheric flight is limited to sub-orbital velocities due to the presence of air, which creates drag. When a satellite enters the atmosphere at orbital velocity, the rate of energy dissipated equals the product of drag, D, and velocity, V. This energy is so high that the surrounding air gets very hot and the satellite is likely to burn up if it does not have thermal protection (e.g., insulating surface treatments or tiles) and thermal management systems on board.

Before moving on to a discussion of the environment within Earth's atmosphere, one last important fact should be noted. In space, interstellar gas molecules are too widely spaced to collide with one another like they do in the Earth's atmosphere. Only by collision can molecules exchange momentum with each to produce pressure, therefore in outer space the pressure equals absolute zero. Lack of collisions eliminates random molecular motions, including rotation and vibration and therefore its internal molecular energy is zero, which corresponds to zero absolute temperature. In summary, in outer space:

$$p_{outer\,space} = 0\,Pa = 0\,psfa \qquad\qquad T_{outer\,space} = 0\,K = 0\,°R$$

3.3   INSIDE THE ATMOSPHERE

Gravitational attraction is the fundamental reason that the gases surrounding the planet are held in place to maintain its life-sustaining atmosphere. It is obvious to any adventurous person that depending upon his or her location within the atmosphere, the quality of the atmospheric air will be very much different. Only a small number of people have ever summited the Earth's highest peak (China/Tibet/Nepal's Mount Everest which is currently approximately 29,035 ft above sea level) without the use of supplementary oxygen bottles to aid in breathing and the use of specially-designed suits to prevent frostbite from exposure to the extreme cold. On the other hand, in the basin of the Earth's lowest point on land (Jordan/Israel's Dead Sea which is about 1,312 ft below sea level) there is no need at all for extra oxygen or warm clothing. Clearly, the atmosphere changes as a function of distance from the Earth. When discussing this distance, one typically refers to the distance as *elevation* if on land, but if one is airborne this distance is typically referred to as *altitude*.

Weather balloons are launched frequently to measure and record (or transmit by radio) the air pressure and temperature as functions of the balloon's altitude. From p and T the air density can be calculated using the equation of state. Obviously readings will be different on hot days and cold days. The barometer reading at sea level is found on the average to equal about 29.92 inHg (760 mmHg). Therefore this was defined as the standard reference pressure called 1 atmosphere (atm). The way a barometer works is as follows: a glass tube, sealed off at one end, is filled with a heavy liquid such as mercury (Hg) with specific gravity, S.G., of 13.6. Covering the open end with a cup seals off the tube filled with mercury. Next the assembly is turned upside-down. The mercury will run into the cup until the weight of the fluid column equals that of a column of air of equal cross-section. The difference, Δh, between the level in the tube and the level in the cup is about 30 inches. This is called the "barometric reading" and depends on atmospheric conditions and altitude, h_G, above sea level. A schematic representation of a mercury barometer is presented in Figure 3-7.

The barometer reading can be converted to units of pressure quite simply (for reasons explained in Chapter 2). Because there is no air above the mercury in the tube, this space is filled with saturated mercury vapor which has negligible pressure so that $p \approx 0$, just like in outer space. The ambient air pressure pushes down on the mercury surface in the cup, which in turn pushes up on the mercury inside the glass tube. A mercury column of height Δh and unit cross-sectional area has a weight per unit cross sectional area equal that of the ambient pressure.

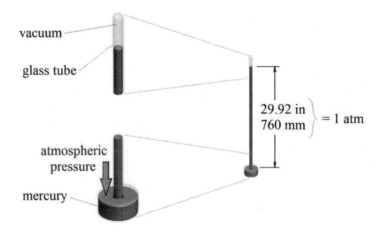

Figure 3-7: A column of mercury contained within a glass tube with its open end kept below the surface of a reservoir of mercury exposed to ambient air provides a method of determining the absolute pressure of the ambient air (higher pressure will cause the column to rise and lower pressure will cause it to fall).

Airplane performance such as cruise speed and fuel consumption rate depends on operational altitude. Performance is thus reported for a specific atmospheric pressure and temperature. To predict aircraft behavior in certain conditions based on the performance specs given, several versions of a "Standard Atmosphere" have been designed and tabulated by various agencies, including the National Air and Space Administration (NASA). In 1959, the US Air Force Air Research and Development Command (ARDC) developed a model to simplify calculations. It agrees reasonably well with worldwide average atmospheric measurements. In many textbooks the ARDC table values are tabulated in both SI and English units.

A slight simplification of the ARDC model has been used here for sample calculations based on constant lapse rates shown in Figure 3-8. This model provides data almost identical to the ARDC model. Both start out at sea level (denoted by subscript "sl") with the internationally agreed upon p_{sl} = 1 atm = 101,325 Pa = 2116 PSFA, temperature T_{sl} = 288.16 K = 518.99 °R and density ρ_{sl} = 1.225 kg/m³ = 0.002377 slug/ft³. The sea level acceleration of gravity is g_o = 9.81 m/s² = 32.2 ft/s² and the Earth's radius is r_e = 6,371 km = 20,902,200 ft.

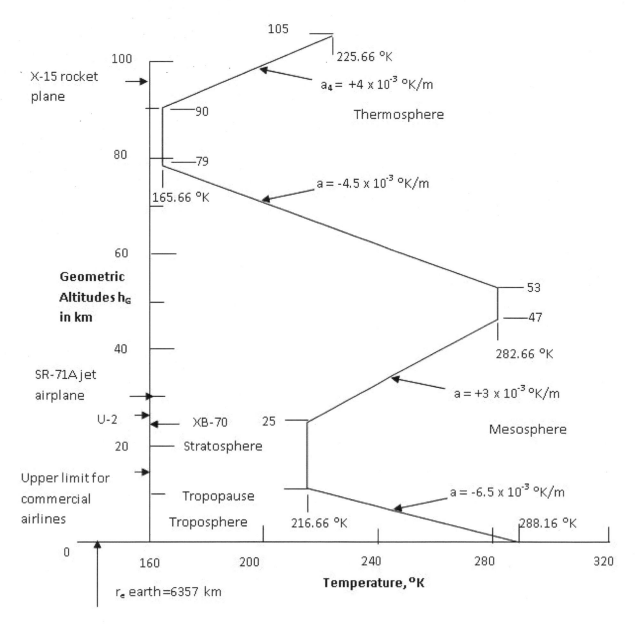

Figure 3-8: "Simplified Standard Atmosphere"- temperature as a function of altitude h_G.

- In the troposphere from sea level at 0 km $< h_G <$ 11 km the temperature drops linearly from 288.16 K at sea level at a lapse rate a = -6.5 x 10^{-3} K/m or a = -6.5 K/km. Note the upper layer of the troposphere is called the tropopause.
- In the stratosphere, from 11 km $< h_G <$ 25 km, the temperature remains constant at 216.66 K.
- In the mesosphere, from 25 km $< h_G <$ 47 km, the temperature rises linearly from 216.66 K to 282.66 K, at a lapse rate a = 3 x 10^{-3} K/m.
- From 47 km $< h_G <$ 53 km, the temperature stays constant at 282.66 K.
- From 53 $< h_G <$ 79 km, the temperature drops linearly from 282.66 K to 165.66 K, at a lapse rate a = -4.5 x 10^{-3} K/m.
- From 79 km to 90 km, the temperature stays constant at 165.66 K.
- The thermosphere is the outer most shell of the atmosphere and starts at 90 km. From there on, the temperature increases steadily.

Note the lapse rate of 6.5 K per 1000 m corresponds to a polytropic expansion of the air with increase in altitude corresponding to k = 1.23, instead of an isentropic expansion with k = 1.4, where k would have been called γ, the specific heat ratio which for air equals 1.4. The lower value of k is caused by heat release from condensing water vapor (note this is the reason for clouds and rain) as air rises in a thermal or thunderstorm. The associated expansion of rising air causes its temperature to drop and when the air temperature drops to the dew point, clouds start to form.

In addition to absolute altitude, h_a, and geometric altitude, h_G, there are many more altitude definitions. One artificial altitude has been introduced to simplify pressure calculations, called the geopotential altitude, h. It is calculated as a function of r_e and h_G as

$$h = h_G \left(\frac{r_e}{r_e + h_G} \right)$$

(3.11)

Admittedly, these three altitudes can cause some confusion among starting engineers. A schematic diagram of the three altitudes, h_a, h_G, and h, is shown in Figure 3-9 to help put them in perspective.

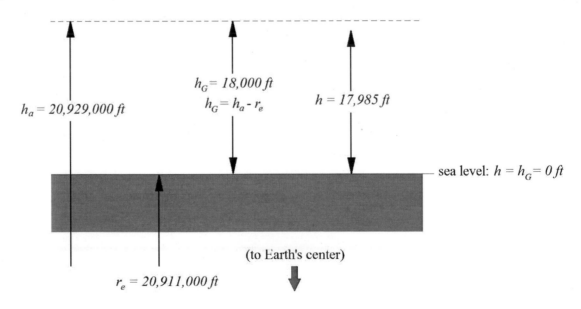

Figure 3-9: Comparison of altitudes: the dashed line is 18,000 ft from sea level, taken by "tape measure".

ENGINEERING IN CODE: TEMPERATURE IN THE EARTH'S ATMOSPHERE

The following MATLAB® script will help the user compute the temperature from 0 to 47,000 m:

```
clear all;
h_G = input('Enter the geometric altitude (in m):   ');
h = h_G*(6357000/(6357000 + h_G));
if h_G <= 11000
    T_ref = 288.16;
    a = -6.5e-3;
    h_ref = 0;
elseif h_G > 11019 & h_G <= 25000
    T_ref = 216.66;
    a = 0;
    h_ref = 11019;
elseif h_G > 25000 & h_G <= 47000
    T_ref = 216.66;
    a = 3e-3;
    h_ref = 25000;
elseif h_G < 0 || h_G > 47000
    disp('Please enter a value between 0 and 47,000 m');
end
T = T_ref + a*(h - h_ref);
disp(sprintf('At h_G = %d, T = %0.2f',h_G,T));
```

Example input and resultant output:
```
Enter the geometric altitude (in m):   8500
At h_G = 8500, T = 232.98
```

Besides those three altitudes, pilots use three additional altitudes in flight which relate altitude, true ambient conditions, and values from the Standard Atmosphere Table

- **Pressure altitude, h_p,** is the altimeter reading obtained with its calibration window adjusted to the "standard day" sea-level value of 29.92 inHg. Alternately, for any specified pressure, read the corresponding "pressure altitude" from the Standard Atmosphere tables. To control traffic at altitudes above 18,000 ft, the FAA uses pressure altitude. For example if FAA traffic control assigns a flight level of 200, then the aircraft is only cleared to operate at a pressure altitude of 20,000 ft. This eliminates the need for the FAA to inform the aircraft of the local sea level barometer reading. The use of flight levels provides safe vertical spacing for commercial high altitude traffic to prevent collisions.
- **Temperature altitude, h_T,** is read from the Standard Atmosphere tables as the altitude where the measured T equals that listed for a standard day. Temperature altitude is important to pilots to identify freezing levels and potentially dangerous icing conditions.
- **Density altitude, h_ρ,** is very important for all flight conditions. It is needed to find from aircraft handbooks what the aircraft performance specifications such as climb rate, flight speed and engine power settings will be. It is also needed to find minimum take-off and landing distances. To determine density altitude, first calculate density from the measured p and T values using the equation of state. Next, look up the corresponding density altitude in the Standard Atmosphere tables.

Figure 3-10 demonstrates how an aircraft may move from one pressure region to another within a non-standard atmosphere without worry that it may cross paths with another aircraft flying in the vicinity. The jet airliner represented in the figure has been instructed to maintain altitude at Federal Aviation Administration (FAA) Flight Level 180, thus the flight crew must fly where the local pressure is equivalent to the pressure of a Standard Atmosphere at 18,000 ft (i.e., p = 1056 psfa). As the airliner flies from the city at location A where the local sea level pressure is relatively high to the city at location C where the local sea level pressure is relatively low it actually must descend in geometric altitude to maintain the same pressure altitude.

On a related note, it is most often the case that where low pressure regions exist, the air is generally moving in an upward direction where it expands as it rises and cools down rapidly. This expansion and cooling process causes much of the gaseous water held within the air to condense, forming clouds and eventually precipitation. The opposite behavior is true in high pressure regions where air masses generally move downward into warmer regions. The compression and heating of the air causes it to act as a sponge, increasing its capacity to hold water vapor and ensuring little or no local precipitation. Again referring to Figure 3-10, it is very likely a nice day for a stroll in City A, but folks are probably going to have to remember to take their umbrellas to work in City C.

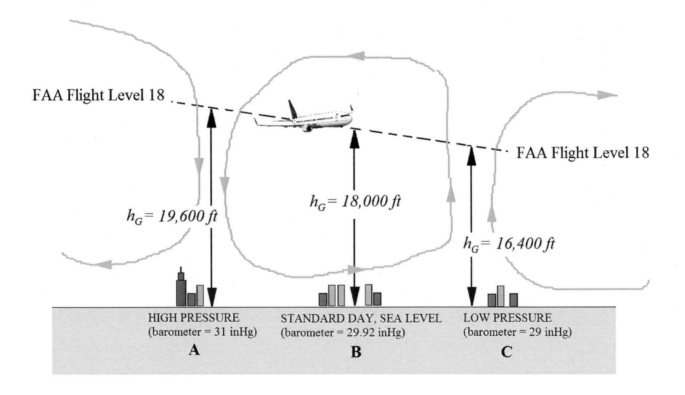

Figure 3-10: Shown here is the effect local sea level barometer reading has on the flight geopotential altitude, when flying at constant p = 1056 psfa of FAA Flight Level 180 through a non-standard atmosphere from a high to low pressure region.

It would appear most useful to be able to calculate the local values for the state variables within the navigable region of the atmosphere to predict potential problems for flight or for a number of other reasons. Table 3-2 presents the basic formulae necessary to calculate these values.

Table 3-2: Procedure for calculating the temperature, pressure and density at any specified geometric altitude from sea level to 47 km.

Description of variables	Sea level (sl) reference values	Equations used for $0\,m < h_G < 11{,}019\,m$ (Troposphere)	Equations used for $11{,}019\,m < h_G < 25{,}000\,m$ (Stratosphere)	Equations used for $25{,}000\,m < h_G < 47{,}000\,m$ (Lower Mesosphere)
Geometric Altitude $h_G\,(m)$	$h_G = h_{G,ref} = 0\,m$	Specify $h_G\,(m)$	Specify $h_G\,(m)$	Specify $h_G\,(m)$
Geopotential Altitude $h\,(m)$	$h = h_{ref} = 0\,m$	$h = h_G\left(\dfrac{r_e}{r_e + h_G}\right)$	$h = h_G\left(\dfrac{r_e}{r_e + h_G}\right)$	$h = h_G\left(\dfrac{r_e}{r_e + h_G}\right)$
Local Gravity $g\,(m/sec^2)$	$g_{sl} = g_0 = 9.81\,m/sec^2$	$g = g_0\left(\dfrac{r_e}{r_e + h_G}\right)^2$	$g = g_0\left(\dfrac{r_e}{r_e + h_G}\right)^2$	$g = g_0\left(\dfrac{r_e}{r_e + h_G}\right)^2$
Lapse Rate $a\,(K/m)$	$a = -0.0065\,(K/m)$	$a = -0.0065\,(K/m)$	$a = 0\,(K/m)$	$a = 0.003\,(K/m)$
Local Temperature $T\,(K)$	$T_{sl} = 288.16\,K$	$T = T_{ref} + a\left(h - h_{ref}\right)$ $T_{ref} = T_{sl}$ $h_{ref} = 0\,m$	$T = T_{ref}$ $T_{ref} = 216.66\,K$ $h_{ref} = 11{,}000\,m$	$T = T_{ref} + a\left(h - h_{ref}\right)$ $T_{ref} = 216.66\,K$ $h_{ref} = 24{,}902\,m$
Local Pressure $p\,(Pa)$	$p_{sl} = 101{,}325\,Pa$	$p = p_{ref}\left(\dfrac{T}{T_{ref}}\right)^{-g_0/(aR)}$ $p_{ref} = p_{sl}$	$p = p_{ref}\,e^{-g_0(h-h_{ref})/(RT_{ref})}$ $p_{ref} = 22{,}616\,Pa$	$p = p_{ref}\left(\dfrac{T}{T_{ref}}\right)^{-g_0/(aR)}$ $p_{ref} = 2{,}523\,Pa$
Local Density $\rho\,(kg/m^3)$	$\rho_{sl} = 1.225\,kg/m^3$	$\rho = \dfrac{p}{RT}$	$\rho = \dfrac{p}{RT}$	$\rho = \dfrac{p}{RT}$

*Gas constant for air is the same for all altitudes, $\quad R_{air} = 287\,N \cdot m/kg \cdot K$

**Radius of the Earth is the same for all altitudes, $\quad r_e = 6{,}357{,}000\,m$

EXAMPLE 3.5: CALCULATING ATMOSPHERIC STATE VARIABLES AT ALTITUDE

Given:

A set of measurements are collected by an aircraft flying in Earth's atmosphere during a standard day at the following geometric altitudes:

$$h_{G,1} = 5,000\,m \qquad h_{G,2} = 15,000\,m \qquad h_{G,3} = 30,000\,m \qquad h_{G,4} = 21,000\,ft$$

Find:

Determine the values of local temperature, pressure, and density at each of the four altitudes (be sure to provide answers in the same unit system with which the altitude is provided).

Solution:

Location 1:

$h_G < 11,019\,m \rightarrow Troposphere$

$a = -0.0065\,K/m, \qquad h_{ref} = 0\,m, \qquad T_{ref} = 288.16\,K, \qquad p_{ref} = 101,325\,Pa$

$$h_1 = h_G\left(\frac{r_e}{r_e + h_G}\right) = 5,000\,m\left(\frac{6371}{6371+5}\right) = 5,000\,m(0.9992) = 4,996\,m$$

$$T_1 = 288.16\,K - 0.0065\,(K/m)(4,996-0)\,m = \boxed{255.69\,K}$$

$$p_1 = 101,325\,Pa\left(\frac{255.69}{288.16}\right)^{-9.81/(-.0065\times287)} = \boxed{54,037\,Pa}$$

$$\rho_1 = \frac{54,037\,Pa}{(287\,N\cdot m/kg\cdot K)(255.69K)} = \boxed{0.7364\,kg/m^3}$$

Location 2:

$11,019\,m < h_G < 25,000\,m \rightarrow Stratosphere$

$a = 0\,K/m, \qquad h_{ref} = 11,000\,m, \qquad T_{ref} = 216.66\,K, \qquad p_{ref} = 22,616\,Pa$

$$h_2 = h_G\left(\frac{r_e}{r_e + h_G}\right) = 15,000\,m\left(\frac{6371}{6371+15}\right) = 15,000\,m(0.9977) = 14,965\,m$$

$$T_2 = T_{ref} = \boxed{216.66\,K}$$

$$p_2 = 22,616\,e^{-9.81(14,965-11,000)/(287\times216.66)} = \boxed{12,099\,Pa}$$

$$\rho_2 = \frac{12,099\,Pa}{(287\,N\cdot m/kg\cdot K)(216.66K)} = \boxed{0.1946\,kg/m^3}$$

Location 3:

$25,000\,m < h_G < 47,000\,m \rightarrow Lower\ Mesosphere$

$a = -0.0065\,K/m, \qquad h_{ref} = 24,902\,m, \qquad T_{ref} = 216.66\,K, \qquad p_{ref} = 2,523\,Pa$

$h_3 = h_G\left(\dfrac{r_e}{r_e + h_G}\right) = 30,000\,m\left(\dfrac{6371}{6371+30}\right) = 30,000\,m(0.9953) = 29,859\,m$

$T_3 = 216.66\,K + 0.003\,(K/m)(29,859 - 24,902)\,m = \boxed{231.53\,K}$

$p_3 = 2,523\,Pa\left(\dfrac{231.52}{216.66}\right)^{-9.81/(0.003\times287)} = \boxed{1,185\,Pa}$

$\rho_3 = \dfrac{1,185\,Pa}{(287\,N\cdot m/kg\cdot K)(231.53\,K)} = \boxed{0.01783\,kg/m^3}$

Location 4:

$h_G = 21,000\,ft \times \left(\dfrac{0.3048\,m}{ft}\right) = 6,401\,m < 11,019\,m \rightarrow Troposphere$

$a = -0.0065\,K/m\left(\dfrac{9}{5}\right)\left(\dfrac{^\circ R}{K}\right)\left(\dfrac{0.3048\,m}{ft}\right) = -0.00357\,^\circ R/ft,$

$h_{ref} = 0\,ft, \qquad T_{ref} = 288.16\,K = 518.7\,^\circ R, \qquad p_{ref} = 101,325\,Pa = 2,116\,psfa$

$r_e = \dfrac{6,371,000\,m}{(0.3048\,m/ft)} = 20,902,223\,ft$

$h_4 = h_G\left(\dfrac{r_e}{r_e + h_G}\right) = 21,000\,ft\left(\dfrac{20,902,223}{20,902,223+21,000}\right) = 21,000\,ft(0.9990) = 20,979\,ft$

$T_4 = 518.7\,^\circ R - 0.00357\,(^\circ R/ft)(20,979 - 0)\,ft = \boxed{443.89\,^\circ R}$

$p_4 = 2,116\,Pa\left(\dfrac{443.89}{518.7}\right)^{-32.2/(-.00357\times1716)} = \boxed{933.2\,psfa}$

$\rho_4 = \dfrac{933.2\,psfa}{(1716\,ft\cdot lb_f/slug\cdot^\circ R)(443.89\,^\circ R)} = \boxed{0.001225\,slug/ft^3}$

Check:

The numerical values and the units appear to be reasonable and address the original question(s).

It is obvious from the recent example calculations for state variables within a Standard Atmosphere that a simple table which included local pressure, temperature, density and more as a function of geometric altitude should be extremely helpful for many engineering applications. In 1959, the US Air Force Air Research and Development Command (ARDC) completed what is now referred to as the ARDC 1959 Model Standard Atmosphere. The model was based on numerous atmospheric measurements taken primarily by rockets (but also by aircraft, balloons and even satellites) ranging in altitudes up to 700 km. It is interesting to note that many of the temperature measurements at extreme altitudes were indirectly measured using the detected local speed of sound. The connection between the speed of sound in a gas and its temperature will be discussed in much more detail later in this book. Some of the assumptions used to model the ARDC Model Standard Atmosphere are as follows:

1) The atmosphere is assumed to obey the ideal gas law, and in particular the state equation

$$\rho = \frac{p}{RT}$$

2) The universal gas constant is taken to be

$$R_u = 8314.39 \, N \cdot m / kmol \cdot K$$

3) The air is assumed to be in hydrostatic equilibrium.
4) Geopotential energy is based on sea level gravitational acceleration as the constant of proportionality, where

$$g_0 = 9.80665 \, m / sec^2$$

5) The value of the Earth's radius taken at a known latitude is

$$r_e = 6,356,766 \, m \qquad \text{at latitude } 45°32'33''$$

6) The molecular-scale temperature of air at altitude varies directly with the absolute temperature of air at altitude by a factor of the ratio of the sea level-to-altitude molecular weight of air, or

$$T_{mol}\left(h_G\right) = T\left(h_G\right) \qquad\qquad for \, h_G < 90 \, km$$

$$T_{mol}\left(h_G\right) = \frac{M_{air,sl}}{M_{air}\left(h_G\right)} T\left(h_G\right) \qquad\qquad for \, h_G \geq 90 \, km$$

7) Based on the assumption in 4), it is possible to express molecular-scale temperature at altitude in terms of the measured molecular-scale gradient (aka, "lapse rate") such that

$$T_{mol}\left(h\right) = T_{mol,base} + L_{mol}\left(h - h_{base}\right) \qquad \begin{cases} base \to \text{value at base of particular layer} \\ L \to \text{molecular-scale gradient, aka lapse rate} \end{cases}$$

8) The standard pressure at sea level is

$$p_{sl} = 101,325 \, Pa = 760 \, mmHg \qquad\qquad \left(\rho_{Hg} = 13595.1 \, kg / m^3\right)$$

9) Pressure-Altitude formulae are

$$p = p_{base} \left[\frac{T_{mol,base}}{T_{mol,base} + L_{mol}\left(h - h_{base}\right)} \right]^{g_0/(RL_{mol})} \qquad for\ L_{mol} \neq 0$$

$$p = p_{base} \exp\left[\frac{-g_0\left(h - h_{base}\right)}{RT_{mol,base}} \right] \qquad for\ L_{mol} = 0$$

10) The atmosphere is assumed to be dry and has a known sea level composition (Table 3-3) and sea level molecular weight,

$$M_{air,sl} = 28.966\ kg/kmol$$

Table 3-3 presents the basic findings for the composition of the Standard Atmosphere, according to USAF ARDC.

Table 3-3: Sea Level atmospheric composition for a dry atmosphere, ARDC Standard Atmosphere, 1959

Constituent Gas	Mole Fraction (%)	Molecular Weight (kg/kmol)
Nitrogen (N_2)	78.09	28.016
Oxygen (O_2)	20.95	32.0000
Argon (A)	0.93	39.944
Carbon dioxide (CO_2)	0.03	44.010
Neon (Ne)	1.8×10^{-3}	20.183
Helium (He)	5.24×10^{-4}	4.003
Krypton (Kr)	1.0×10^{-4}	83.7
Hydrogen (H_2)	5.0×10^{-5}	2.0160
Xenon (Xe)	8.0×10^{-6}	131.3
Ozone (O_3)	1.0×10^{-6}	48.0000
Radon (Rn)	6.0×10^{-18}	222.0

In the following section, a simplified set of Standard Atmosphere tables (one in SI and one in English units) is presented as a handy reference for the reader. Calculations to build these tables have been based purely on the formulae provided in Table 3-2 and associated error percentages with respect to the more accurate ARDC tables are presented as footnotes.

3.4   SIMPLIFIED STANDARD ATMOSPHERE TABLES

Table 3-4: Simplified Standard Atmosphere Table (SI)

Metric Version

For ARDC model see "Air Force Cambridge Research Center Report" No: TR-59-267

Reference values used:

Earth radius r_e = 6,357 km, sea level gravity g_o = 9.81 m/sec^2, gas constant R = 287 N·m/kg·K

From sea level, geometric altitude h_G = 0 m to 11019 m and equipotential reference altitude h_{ref} = 0 Use: T_{ref} = 288.16 K, p_{ref} = 101,325 Pa, h_{ref} = 0 m and $\rho = p/(RT)$ or ρ_{ref} = 1.225 kg/m^3 with constant lapse rate $dT/dh = a = -0.0065$ K/m to altitude h_G = 11,019 or h = 11,000 m

From h_G = 11,019 m to 25,000 m change lapse rate to: a = 0, with T constant = 216.66 K and at h_G = 25,000 find p = 2,523 Pa and T = 216.66 K

From h_G = 25,000 m to 47,000 m change lapse rate to: a = +0.003 K/m and use p_{ref} = 2,523 Pa and T_{ref} = 216.66 K and h_{ref} = 25,000 m.

Acceleration of gravity decreases with the square of the radial distance $g = g_o[r/(r+h_G)]^2$

geometric altitude, h_G (m)	temperature, T (K)	lapse rate, dT/dh (K/m)	pressure, p (Pa)	density, ρ (kg/m^3)	acceleration gravity, g (m/sec^2)	geopotential altitude, h (m)
0	288.16	-0.0065	101,325	1.2250	9.8100	0
500	284.91	-0.0065	95,458	1.1674	9.8085	500
1000	281.66	-0.0065	89,871	1.1118	9.8069	1,000
1500	278.41	-0.0065	84,552	1.0582	9.8054	1,500
2000	275.16	-0.0065	79,492	1.0066	9.8038	1,999
2500	271.92	-0.0065	74,681	0.9570	9.8023	2,499
3000	268.67	-0.0065	70,108	0.9092	9.8007	2,999
3500	265.42	-0.0065	65,766	0.8633	9.7992	3,498
4000	262.18	-0.0065	61,645	0.8193	9.7977	3,997
4500	258.93	-0.0065	57,737	0.7769	9.7961	4,497
5000	255.69	-0.0065	54,032	0.7363	9.7946	4,996
5500	252.44	-0.0065	50,522	0.6973	9.7930	5,495
6000	249.20	-0.0065	47,200	0.6600	9.7915	5,994
6500	245.95	-0.0065	44,058	0.6241	9.7900	6,493
7000	242.71	-0.0065	41,087	0.5898	9.7884	6,992
7500	239.47	-0.0065	38,281	0.5570	9.7869	7,491
8000	236.23	-0.0065	35,633	0.5256	9.7854	7,990
8500	232.98	-0.0065	33,136	0.4956	9.7838	8,489
9000	229.74	-0.0065	30,783	0.4669	9.7823	8,987
9500	226.50	-0.0065	28,567	0.4395	9.7807	9,486
10000	223.26	-0.0065	26,483	0.4133	9.7792	9,984
10500	220.02	-0.0065	24,523	0.3884	9.7777	10,483
11000	216.78	-0.0065	22,683	0.3646	9.7761	10,981
11019	216.66	-0.0065	22,616	0.3637	9.7761	11,000

geometric altitude, h_G (m)	temperature, T (K)	lapse rate, dT/dh (K/m)	pressure, p (Pa)	density, ρ (kg/m^3)	acceleration gravity, g (m/sec^2)	geopotential altitude, h (m)
11500	216.66	0	20,969	0.3372	9.7746	11,479
12000	216.66	0	19,384	0.3117	9.7731	11,977
12500	216.66	0	17,919	0.2882	9.7715	12,475
13000	216.66	0	16,565	0.2664	9.7700	12,973
13500	216.66	0	15,314	0.2463	9.7685	13,471
14000	216.66	0	14,157	0.2277	9.7669	13,969
14500	216.66	0	13,088	0.2105	9.7654	14,467
15000	216.66	0	12,099	0.1946	9.7639	14,965
15500	216.66	0	11,186	0.1799	9.7623	15,462
16000	216.66	0	10,341	0.1663	9.7608	15,960
16500	216.66	0	9,561	0.1538	9.7593	16,457
17000	216.66	0	8,839	0.1422	9.7577	16,955
17500	216.66	0	8,172	0.1314	9.7562	17,452
18000	216.66	0	7,556	0.1215	9.7547	17,949
18500	216.66	0	6,986	0.1123	9.7532	18,446
19000	216.66	0	6,459	0.1039	9.7516	18,943
19500	216.66	0	5,972	0.0960	9.7501	19,440
20000	216.66	0	5,522	0.0888	9.7486	19,937
20500	216.66	0	5,105	0.0821	9.7470	20,434
21000	216.66	0	4,720	0.0759	9.7455	20,931
21500	216.66	0	4,365	0.07019	9.7440	21,428
22000	216.66	0	4,036	0.06490	9.7425	21,924
22500	216.66	0	3,732	0.06001	9.7409	22,421
23000	216.66	0	3,451	0.05549	9.7394	22,917
23500	216.66	0	3,191	0.05131	9.7379	23,413
24000	216.66	0	2,950	0.04745	9.7363	23,910
24500	216.66	0	2,728	0.04388	9.7348	24,406
25000	216.66	0	2,523	0.04057	9.7333	24,902
25500	218.15	0.003	2,334	0.03727	9.7318	25,398
26000	219.64	0.003	2,160	0.03426	9.7302	25,894
26500	221.12	0.003	2,000	0.03151	9.7287	26,390
27000	222.61	0.003	1,853	0.02900	9.7272	26,886
27500	224.10	0.003	1,717	0.02670	9.7257	27,382
28000	225.59	0.003	1,593	0.02460	9.7241	27,877
28500	227.07	0.003	1,478	0.02268	9.7226	28,373
29000	228.56	0.003	1,372	0.02092	9.7211	28,868
29500	230.05	0.003	1,274	0.01930	9.7196	29,364
30000	231.53	0.003	1,184	0.01782	9.7181	29,859

Accuracy compared with ARDC table:

at h_G = 0 m , find error in T = 0%, error in p =0%, error in ρ = 0%

at h_G = 11,000 m , find error in T = 0%, error in p =0.07%, error in ρ = 0.05%

at h_G = 25,000 m , find error in T = 0%, error in p =0.16%, error in ρ = 0.17%

at h_G = 30,000 m , find error in T = 0.125%, error in p =0.08%, error in ρ = 0.2%

Table 3-5: Simplified Standard Atmosphere Table (English)

English Version						
For ARDC model see "Air Force Cambridge Research Center Report" No: TR-59-267						
Reference values used:						
Earth radius r_e = 20,910,000 ft, sea level gravity g_o = 32.2 ft/sec^2, gas constant R = 1,716 ft·lb$_f$/slug·°R						
From sea level h_G = 0 to h_G = 36,065 ft geometric altitude, use h_{ref} = $h_G[r/(r+h_G)]$ = 0 ft with T_{ref} = 518.7 °R, p_{ref} = 2116 psfa and ρ = $p_{ref}/(RT_{ref})$ = 0.002377 slug/ft^3 and constant lapse rate dT/dh_G = a = - 0.003575 °R/ft to altitude h_G = 36,065 ft where T = 389.99 °R						
From h_G = 36,065 ft to 82,350 ft change lapse rate to: a = 0, then T is constant = 389.99 °R and p_{ref} = 474 Pa and h_{ref} = 36003 (ft)						
From h_G = 82,350 ft to 154,000 ft change lapse rate to: a = +0.00164 with p_{ref} = 52 Pa and T_{ref} = 389.99 °R with h_{ref}= 82027 ft.						
Acceleration of gravity decreases with the square of the radial distance g = $g_o[r/(r+h_G)]^2$						
geometric altitude, h_G (ft)	temperature, T (°R)	lapse rate, dT/dh (°R/ft)	pressure, p (psfa)	density, ρ (slug/ft^3)	acceleration gravity, g (ft/sec^2)	geopotential altitude, h (ft)
0	518.7	-0.003575	2,116	0.0023770	32.2000	0
1000	515.13	-0.003575	2,041	0.0023085	32.1969	1,000
2000	511.55	-0.003575	1,967	0.0022412	32.1938	2,000
3000	507.98	-0.003575	1,896	0.0021754	32.1908	3,000
4000	504.40	-0.003575	1,827	0.0021111	32.1877	3,999
5000	500.83	-0.003575	1,760	0.0020483	32.1846	4,999
6000	497.26	-0.003575	1,695	0.0019869	32.1815	5,998
7000	493.68	-0.003575	1,632	0.0019269	32.1785	6,998
8000	490.11	-0.003575	1,571	0.0018684	32.1754	7,997
9000	486.54	-0.003575	1,512	0.0018112	32.1723	8,996
10000	482.97	-0.003575	1,455	0.0017554	32.1692	9,995
11000	479.40	-0.003575	1,399	0.0017009	32.1661	10,994
12000	475.82	-0.003575	1,345	0.0016477	32.1631	11,993
13000	472.25	-0.003575	1,293	0.0015958	32.1600	12,992
14000	468.68	-0.003575	1,243	0.0015452	32.1569	13,991
15000	465.11	-0.003575	1,194	0.0014958	32.1539	14,989
16000	461.54	-0.003575	1,147	0.0014476	32.1508	15,988
17000	457.97	-0.003575	1,101	0.0014006	32.1477	16,986
18000	454.41	-0.003575	1,056	0.0013548	32.1446	17,985
19000	450.84	-0.003575	1,014	0.0013102	32.1416	18,983
20000	447.27	-0.003575	972	0.0012667	32.1385	19,981
21000	443.70	-0.003575	932	0.0012243	32.1354	20,979
22000	440.13	-0.003575	894	0.0011830	32.1323	21,977
23000	436.57	-0.003575	856	0.0011428	32.1293	22,975
24000	433.00	-0.003575	820	0.0011037	32.1262	23,972
25000	429.43	-0.003575	785	0.0010656	32.1231	24,970
26000	425.87	-0.003575	752	0.0010285	32.1201	25,968

geometric altitude, h_G (ft)	temperature, T (°R)	lapse rate, dT/dh (°R/ft)	pressure, p (psfa)	density, ρ (slug/ft³)	acceleration gravity, g (ft/sec²)	geopotential altitude, h (ft)
27000	422.30	-0.003575	719	0.0009924	32.1170	26,965
28000	418.73	-0.003575	688	0.0009573	32.1139	27,963
29000	415.17	-0.003575	658	0.0009231	32.1109	28,960
30000	411.60	-0.003575	629	0.0008899	32.1078	29,957
31000	408.04	-0.003575	600	0.0008576	32.1047	30,954
32000	404.47	-0.003575	573	0.0008262	32.1017	31,951
33000	400.91	-0.003575	547	0.0007957	32.0986	32,948
34000	397.35	-0.003575	522	0.0007661	32.0955	33,945
35000	393.78	-0.003575	498	0.0007374	32.0925	34,942
36000	390.22	-0.003575	475	0.0007094	32.0894	35,938
36065	389.99	-0.003575	474	0.0007076	32.0892	36,003
37000	389.99	0	454	0.0006787	32.0863	36,935
38000	389.99	0	433	0.0006470	32.0833	37,931
39000	389.99	0	413	0.0006167	32.0802	38,927
40000	389.99	0	393	0.0005879	32.0772	39,924
41000	389.99	0	375	0.0005604	32.0741	40,920
42000	389.99	0	357	0.0005342	32.0710	41,916
43000	389.99	0	341	0.0005092	32.0680	42,912
44000	389.99	0	325	0.0004854	32.0649	43,908
45000	389.99	0	310	0.0004627	32.0619	44,903
46000	389.99	0	295	0.0004411	32.0588	45,899
47000	389.99	0	281	0.0004204	32.0557	46,895
48000	389.99	0	268	0.0004008	32.0527	47,890
49000	389.99	0	256	0.0003821	32.0496	48,885
50000	389.99	0	244	0.0003642	32.0466	49,881
51000	389.99	0	232	0.0003472	32.0435	50,876
52000	389.99	0	221	0.0003310	32.0404	51,871
53000	389.99	0	211	0.0003155	32.0374	52,866
54000	389.99	0	201	0.0003008	32.0343	53,861
55000	389.99	0	192	0.0002867	32.0313	54,856
56000	389.99	0	183	0.0002733	32.0282	55,850
57000	389.99	0	174	0.0002606	32.0252	56,845
58000	389.99	0	166	0.0002484	32.0221	57,840
59000	389.99	0	158	0.0002368	32.0191	58,834
60000	389.99	0	151	0.0002257	32.0160	59,828
61000	389.99	0	144	0.0002152	32.0129	60,823
62000	389.99	0	137	0.0002051	32.0099	61,817

Accuracy compared with ARDC table:

at h_G = 0 ft , find error in T = 0%, error in p =0%, error in ρ = 0%

at h_G = 37,000 ft , find error in T = 0%, error in p =0%, error in ρ = 0.1%

at h_G = 82,000 ft , find error in T = 0%, error in p =0.3%, error in ρ = 0%

at h_G = 100,000 ft , find error in T = 0.03%, error in p =0%, error in ρ = 1.5%

geometric altitude, h_G (ft)	temperature, T (°R)	lapse rate, dT/dh (°R/ft)	pressure, p (psfa)	density, ρ (slug/ft^3)	acceleration gravity, g (ft/sec^2)	geopotential altitude, h (ft)
63000	389.99	0	131	0.0001956	32.0068	62,811
64000	389.99	0	125	0.0001864	32.0038	63,805
65000	389.99	0	119	0.0001777	32.0007	64,799
66000	389.99	0	113	0.0001694	31.9977	65,792
67000	389.99	0	108	0.0001615	31.9946	66,786
68000	389.99	0	103	0.0001540	31.9916	67,780
69000	389.99	0	98	0.0001468	31.9885	68,773
70000	389.99	0	94	0.0001400	31.9855	69,766
71000	389.99	0	89	0.0001334	31.9824	70,760
72000	389.99	0	85	0.0001272	31.9794	71,753
73000	389.99	0	81	0.0001213	31.9763	72,746
74000	389.99	0	77	0.0001156	31.9733	73,739
75000	389.99	0	74	0.0001102	31.9702	74,732
76000	389.99	0	70	0.0001051	31.9672	75,725
77000	389.99	0	67	0.0001002	31.9642	76,717
78000	389.99	0	64	0.0000955	31.9611	77,710
79000	389.99	0	61	0.0000911	31.9581	78,703
80000	389.99	0	58	0.0000868	31.9550	79,695
81000	389.99	0	55	0.0000828	31.9520	80,687
82000	389.99	0	53	0.0000789	31.9489	81,680
82350	389.99	0	52	0.0000776	31.9479	82,027
83000	391.05	0.00164	51	0.0000763	31.9459	82,672
84000	392.67	0.00164	49	0.0000725	31.9428	83,664
85000	394.30	0.00164	47	0.0000688	31.9398	84,656
86000	395.93	0.00164	44	0.0000654	31.9368	85,648
87000	397.55	0.00164	42	0.0000622	31.9337	86,640
88000	399.18	0.00164	40	0.0000591	31.9307	87,631
89000	400.81	0.00164	39	0.0000562	31.9276	88,623
90000	402.43	0.00164	37	0.0000534	31.9246	89,614
91000	404.06	0.00164	35	0.0000508	31.9216	90,606
92000	405.68	0.00164	34	0.0000483	31.9185	91,597
93000	407.31	0.00164	32	0.0000460	31.9155	92,588
94000	408.94	0.00164	31	0.0000437	31.9124	93,579
95000	410.56	0.00164	29	0.0000416	31.9094	94,570
96000	412.19	0.00164	28	0.0000396	31.9064	95,561
97000	413.81	0.00164	27	0.0000377	31.9033	96,552
98000	415.44	0.00164	26	0.0000360	31.9003	97,543
99000	417.06	0.00164	25	0.0000342	31.8972	98,533
100000	418.68	0.00164	23	0.0000326	31.8942	99,524

3.5   DERIVATION OF GOVERNING EQUATIONS INSIDE THE ATMOSPHERE

GEOPOTENTIAL ALTITUDE

For an atmosphere, pressure varies with geometric altitude as

$$\frac{dp}{dh_G} = -\rho g$$

Assume some altitude modification such that g will always be constant and equal to g_0. Call this geopotential altitude, h, so that now

$$\frac{dp}{dh} = -\rho g_0$$

One can now write

$$\frac{dp}{\rho} = -g_0 dh = -g dh_G$$

Or

$$dh = \left(\frac{g}{g_0}\right) dh_G = \left(\frac{r_e}{r_e + h_G}\right)^2 dh_G$$

Now it is necessary to integrate from 0 to h or h_G:

$$\int_0^h dh = \int_0^{h_G}\left[\frac{r_e^2}{\left(r_e + h_G\right)^2}\right] dh_G = r_e^2 \int_0^{h_G}\left[\frac{1}{\left(r_e + h_G\right)^2}\right] dh_G$$

From standard integral tables, find that

$$\int \frac{1}{\left(x+a\right)^2} dx = -\frac{1}{\left(x+a\right)} + c$$

So the RHS of the integration will be

$$r_e^2 \int_0^{h_G}\left[\frac{1}{\left(r_e + h_G\right)^2}\right] dh_G = -r_e^2\left[\frac{1}{\left(r_e + h_G\right)}\right]_0^{h_G} = -r_e^2\left[\frac{1}{r_e + h_G} - \frac{1}{r_e}\right] = -r_e^2\left[\frac{r_e}{r_e\left(r_e + h_G\right)} - \frac{\left(r_e + h_G\right)}{r_e\left(r_e + h_G\right)}\right]$$

$$= -r_e^2\left[\frac{-h_G}{r_e\left(r_e + h_G\right)}\right] = r_e\left(\frac{h_G}{r_e + h_G}\right) = h_G\left(\frac{r_e}{r_e + h_G}\right)$$

The LHS of the integration is easy:

$$\int_0^h dh = [h]_0^h = (h-0) = h$$

So finally, one may arrive at an expression for geopotential altitude, finding that

$$h = h_G \left(\frac{r_e}{r_e + h_G} \right)$$

TEMPERATURE

Temperature within the atmosphere is quite easy to calculate with a known geopotential altitude. Keep in mind that lapse rate is just a way of describing the linear relationship between temperature and altitude. In other words,

$$a = \frac{dT}{dh}$$

A straightforward integration will yield the equation for temperature,

$$T = T_{ref} + a\left(h - h_{ref}\right)$$

PRESSURE

Recall that an effort to solve for some new way of expressing altitude so that gravitational acceleration remained constant made it possible to substitute g_0 and h into equations which previously used g and h_G. Quickly look at what that produces:

$$\frac{dp}{dh} = -\rho g_0 \quad \Rightarrow \quad dp = -\rho g_0 dh = -\left(\frac{p}{RT} \right) g_0 dh$$

So

$$\frac{dp}{p} = -\frac{g_0}{R}\frac{dh}{T}$$

Recognizing the expression for lapse rate, one may note that for layers with constant, non-zero lapse rates

$$a = \frac{dT}{dh} \quad \Rightarrow \quad dh = \frac{dT}{a}$$

Which means that

$$\frac{dp}{p} = -\frac{g_0}{aR}\frac{dT}{T}$$

Integrating the equation between the reference condition and the current value,

$$\int_{P_{ref}}^{p}\frac{dp}{p} = -\left(\frac{g_0}{aR}\right)\int_{T_{ref}}^{T}\frac{dT}{T} \qquad \Rightarrow \ln\left(\frac{p}{p_{ref}}\right) = -\left(\frac{g_0}{aR}\right)\ln\left(\frac{T}{T_{ref}}\right) \qquad \Rightarrow e^{\ln\left(\frac{p}{p_{ref}}\right)} = e^{\ln\left(\left(\frac{T}{T_{ref}}\right)^{-\left(\frac{g_0}{aR}\right)}\right)}$$

And finally arrive at an expression for pressure within a layer of constant, non-zero lapse rate:

$$\boxed{\frac{p}{p_{ref}} = \left(\frac{T}{T_{ref}}\right)^{-\left(\frac{g_0}{aR}\right)}} \qquad a \neq 0$$

What about in the case of a zero lapse rate layer like the Earth's stratosphere? Using an earlier equation but considering the fact that temperature is constant:

$$\frac{dp}{p} = -\frac{g_0}{R}\frac{dh}{T} = -\frac{g_0}{RT_{ref}}dh$$

Now integrate

$$\int_{P_{ref}}^{p}\frac{dp}{p} = -\frac{g_0}{RT_{ref}}\int_{h_{ref}}^{h}dh \qquad \Rightarrow \ln\left(\frac{p}{p_{ref}}\right) = -\left(\frac{g_0}{RT_{ref}}\right)(h - h_{ref}) \qquad \Rightarrow e^{\ln\left(\frac{p}{p_{ref}}\right)} = e^{-\left(\frac{g_0}{RT_{ref}}\right)(h - h_{ref})}$$

And finally arrive at an expression for pressure within a layer of constant temperature:

$$\boxed{\frac{p}{p_{ref}} = e^{-\left[\frac{g_0(h - h_{ref})}{RT_{ref}}\right]}} \qquad a = 0$$

3.6 SUMMARY

OUTSIDE THE ATMOSPHERE

Gravitational acceleration between two masses, m_1 and m_2, yields an attractive force, F, such that

$$F = G\frac{m_1 m_2}{r^2}$$

where

$$G = 6.674 \times 10^{-11} \; N \cdot m^2/kg^2$$

Escape velocity from massive objects is calculated as

$$V_e = \sqrt{\frac{2mG}{r}}$$

where m is the mass of the massive object.

For orbital mechanics, where

$$r_{orbit} = r_p + h_G = h_a$$

when a satellite is in equilibrium, the gravitational acceleration must be equal in magnitude and opposite in direction to the centripetal acceleration, such that

$$g = g_0 \left(\frac{r_p}{r_{orbit}}\right)^2 = a_c = \frac{V^2_{orbit}}{r_{orbit}}$$

From this relationship, the orbital velocity is found as

$$V_{orbit} = \sqrt{g r_{orbit}} = \sqrt{g_0 \left(\frac{r_p^2}{r_{orbit}}\right)}$$

and the period of the satellite as

$$t = \frac{\Delta d}{V} = \frac{2\pi r_{orbit}}{V_{orbit}}$$

Conic sections are representative of the paths celestial bodies may take when moving near one another. The four primary paths and some associated details are:

Section	Eccentricity	Equation
Circular	$e = 0$	$x^2 + y^2 = r^2$
Elliptical	$0 < e = \sqrt{1 - \left(\frac{b}{a}\right)^2} < 1$	$\left(\frac{x}{a}\right)^2 + \left(\frac{y}{b}\right)^2 = 1$
Parabolic	$e = 1$	$y^2 = 4px$
Hyperbolic	$e = \sqrt{1 + \left(\frac{b}{a}\right)^2} > 1$	$\left(\frac{x}{a}\right)^2 - \left(\frac{y}{b}\right)^2 = 1$

INSIDE THE ATMOSPHERE

Gravity holds gases of an atmosphere close to the surface. Different lapse rates define the atmospheric regions that exist.

Lapse rate	Layer	h_G (m)
	exosphere	>350,000
$a > 0$	thermosphere	90,000 - 350,000
$a > 0, a = 0, a < 0$	mesosphere	25,000 - 90,000
$a = 0$	stratosphere	11,019 - 25,000
$a < 0$	troposphere	0 - 11,019

Geopotential altitude is designed to force a constant gravitational field within the atmosphere, and is defined as

$$h = h_G \left(\frac{r_p}{h_G + r_p} \right)$$

For zones with a constant non-zero lapse rate,

$$T = T_{ref} + a \left(h - h_{ref} \right) \qquad p = p_{ref} \left(\frac{T}{T_{ref}} \right)^{-\frac{g_0}{aR}} \qquad \rho = \frac{p}{RT}$$

For zones with a constant lapse rate of zero,

$$T = T_{ref} \qquad p = p_{ref} e^{-\frac{g_0 \left(h - h_{ref} \right)}{RT_{ref}}} \qquad \rho = \frac{p}{RT}$$

CHAPTER 3 PROBLEMS

3.1) Assume the Sun and the Earth are point masses and that the Earth has a circular orbit about the Sun. Use Eq. (3.1), Eq. (3.7), Eq. (3.9), and Table 3-1 to approximate the orbital radius of the Earth. Be sure to document each step in the solution process.

3.2) A spaceship enters into a circular orbit around some planet X, at an orbit radius r_{orbit} = 10,000 km. Planet X has a surface radius r_{planet} = 4,000 km and surface gravity g_o=3m/s²

    a. Calculate the acceleration of gravity at the altitude of the orbit radius r_{orbit}.
    b. Calculate spaceship circular orbital velocity V_{orbit} in m/s and orbit time t_{orbit} in hrs.

3.3) The fictional planet Plioerus (with radius r = 5000 km), has a hydrogen atmosphere with R = 4157 N·m/kg·K and *uniform* temperature T_o = -100 °C. On the surface the acceleration of gravity g_o = 6.42 m/s² and the gas pressure p_{ref} = 52,300 Pa.

    a. Calculate the density of the hydrogen gas at the surface in kg/m³.
    b. Calculate the pressure of the hydrogen at a geopotential altitude h = 20 km
    c. Calculate the density of the hydrogen at a geopotential altitude h = 20 km
    d. Calculate the gravity g at a geopotential altitude h = 20 km
    e. Calculate the geometric altitude h_G at a geopotential altitude h = 20 km

3.4) Create a code using MATLAB® that can be used to calculate and display values for p, T, ρ, h, and g on Earth for a user-prompted input value of h_G. Use the code to find the values of p, T, ρ, h, and g at the following geometric altitudes. Include a copy of the script file along with the answers.

    a. h_G = 21,500 ft (report your values in English units)
    b. h_G = 26,500 m (report your values in SI units)

3.5) Use MS Excel® to recreate the Metric version of the simplified ARDC standard atmospheric table, using an incremental change in geometric altitude of 250 m. Print out the entire table, going from a value of h_G from 0 km to 30 km.

3.6) On earth, calculate the values of p, T, ρ and g (in SI units) on a standard day at:

    a. A geometric altitude h_G = 6600 m
    b. A geometric altitude h_G = 36600 m

3.7) An airplane operates in a non-standard atmosphere. Its ambient pressure p = 44,000 Pa and temperature T = -12 °C.

    a. Calculate the local air density in kg/m³.
    b. Read from std. atm. tables its nearest "pressure altitude"
    c. Read from std. atm. tables its nearest "temperature altitude"
    d. Read from std. atm. tables its nearest "density altitude"

3.8) An airplane operates in a non-standard atmosphere. The pilot reads the instruments which indicate the following values: T = 25°F, altimeter reading "pressure altitude" h_p = 12,000 ft.

    a. Find the ambient air pressure in psfa from the tables.
    b. Calculate the ambient air density ρ_{amb} in slug/ft³

CHAPTER 4.   AERODYNAMICS

4.1   INTRODUCTION

The study of fluids with particular emphasis on the forces which they may exert on a system is known as "fluid mechanics". The study of stationary fluids is called "fluid statics", or "hydrostatics", and that of flowing fluids is called "fluid dynamics", or "hydrodynamics". The study of fluid mechanics is a broad topic, and includes the flow of liquids, vapors, gases, and even plasma.

1. Liquids have a temperature $T < T_{boiling}$ and increase little in density with a rise in pressure. When a liquid is heated to the boiling temperature, some of the fluid mass receives enough energy so that it changes on a molecular level from a liquid into a vapor, which then bubbles up out of the liquid because it is less dense. Only when a liquid boils at the critical pressure, $p_{critical}$, does the liquid transform into a vapor without bubbling or increasing in volume. That boiling temperature is called the critical temperature, $T_{critical}$.
2. Vapor escapes from a liquid during boiling. Vapors occur within the temperature range given as $T_{boiling} < T < T_{critical}$. Vapor has enthalpy, h, specific heat, c_p, and density, ρ, that are highly dependent on p and T, so their values are tabulated in thermodynamic tables.
3. Gases have a temperature $T > T_{critical}$. The further the gas temperature, T, is above $T_{critical}$, the less significant the intermolecular forces become. It is therefore considered to behave like an ideal gas. If the gas also has a nearly constant specific heat, c_p, over the range of temperature of interest, it is called a "calorically perfect gas". Assuming a constant specific heat eliminates the need for thermodynamic tables and simplifies specific internal energy to $e = c_v T$ and specific enthalpy to $h = c_p T$. With p and T known, density is best found using the ideal gas equation of state, $\rho = p/(RT)$.

In the study of aircraft performance, the fluid involved is usually just air. In this case, the study of "hydrodynamics" is then called "aerodynamics". The word itself stems from the Greek root words *aer* (meaning "air") and *dynamis* (meaning "power"), indicating that the study of aerodynamics pertains especially to powered air, or more generally speaking, powered gas.

With some mathematical manipulation, the laws of physics can often provide, with amazing accuracy, the fluid pressure, velocity, and temperature distribution on any object immersed in a flowing fluid like water or air. Calculating these parameters has the advantage of being faster, cheaper, and providing more detail than is possible by conducting experimental measurements. Unfortunately viscous effects due to fluid viscosity and its associated shear stress, τ, can add so much complexity to these calculations that assumptions and mathematical simplifications must often be made to proceed. Therefore any numerical solution for a new flight configuration requires some experimental verification to gain confidence in the calculated data. Sometimes the experimental data prove that the computational results are wrong. Such cases call for improvements in the theoretical modeling, often relating to assumptions made in regard to flow phenomena like turbulence and separation. Continuous progress in flow modeling and adaptive grid development is increasing the number of fluid flow problems that can be solved numerically.

For most streamlined objects in flight, viscous effects such as wall friction, flow turbulence and separation can be treated separately as they only affect a thin region around the body called the boundary layer (often abbreviated, BL). For most cases, fluid pressure on the body is equivalent to the pressure just outside the boundary layer. The remaining flow field can be treated as friction-free or "inviscid". The calculated aerodynamic force can afterwards be improved by adding viscous effects such as skin friction drag and pressure drag.

Inviscid flow is covered in the first part of this chapter. This type of flow problem depends only on four flow parameters: V, p, T and ρ. In the second part of this chapter, viscous drag is discussed where the shear stress, τ, cannot be ignored.

Almost all liquids have the characteristic of being nearly incompressible, regardless of ambient conditions. This fact enables actually reduces the complexity of flow calculations because the fluid density is effectively a constant value. Gases, on the other hand, are easily compressed and therefore necessitate the recalculation of density whenever other state variables like pressure and temperature change in a flow. Equations used in the study of compressible flow of an ideal gas are:

1) Equation of State (relates ρ to p and T)
2) Continuity Equation (gives mass flow rate as a function of V and ρ)
3) Momentum Equation for Inviscid Flow (relates p to V and ρ)
4) Energy Equation (relates T to V)
5) Isentropic Equation for Inviscid Flow (relates T to p)
6) Bernoulli Equation for incompressible flow (relates p and V)

4.2   STEADY AND UNSTEADY FLOWS

The flow field around an airplane is generated by pressure and rarefaction waves (acoustic waves) emitted from its surface. These waves travel in all directions at the local speed of sound, a. These waves generate pressure gradients within the flow field, which are necessary to accelerate or decelerate the fluid in the proper direction in order to allow the airplane to pass through. When one plays music, acoustic waves are continuously generated by strings (e.g., guitar strings or piano wires) or by a vibrating surface (e.g., the batter head on a drum). The associated back and forth airflow generated can be felt by holding one's hand in front of a speaker. Although the flow may be periodic (i.e., has a definite repeating pattern over time), such airflow is obviously "unsteady" as it does not remain constant in time. On the other hand when one watches the flow in a slow river, the flow appears "steady", meaning that at any point (x, y, z) in the water, the velocity remains the same in time. Even for fast flowing turbulent rivers, one considers the flow to be steady because the time-averaged velocity at a point does not change with time.

In comparison, when standing on the sidewalk watching a car go by, one always witnesses an unsteady flow field. There will be a momentary rush of wind generated by the car as it goes by. When it snows, one can observe the sudden interruption in the path of the snowflakes when the car passes by. Moments before the car arrives, the snowflakes are seen being pushed out of the way to let the car go by. When the snowflakes arrive at the back of the car, they tend to follow the car in its wake before settling down to the ground. The light snow flakes, moving with the local air velocity, show the necessary change in air velocity at any point (x, y, z), as a function of time. Therefore the flow field generated by the car as it appears to a fixed observer on the side of the road is unsteady. Note if a butterfly happens to fly in the path of the car, one can see how the flow field induced by the presence of the car first pushes the butterfly forwards and up above the hood and windshield of the car. When reaching the top of the car, the butterfly speeds up to a point beyond its original position. Behind the car, the butterfly moves down and forward to return to its original position. The butterfly thus performs a high speed loop de loop by riding within the motion of its home fluid.

Mathematically, steady and unsteady flow fields may be identified quite easily. It should be apparent from the last few examples that the only difference between the two cases is the question of whether any variables within the fluid were functions of time (i.e., whether variables at a given location changed from one moment to the next). The time derivatives of any of the fluid's parameters (T, p, ρ, V, τ, etc.) will immediately indicate a flow field's steady or unsteady nature such that

$$\text{"steady"} \Rightarrow \quad \frac{\partial}{\partial t}(\ ) \equiv 0$$

$$\text{"unsteady"} \Rightarrow \frac{\partial}{\partial t}(\ ) \neq 0$$

Consider the case of the unsteady flow field around the car once again from a different perspective. As a driver of the car, one sees a totally different flow field. The snowflakes or butterflies appear to come towards the car from infinitely far ahead, with a velocity V_∞. But just before impact, they smoothly accelerate up and over the hood and windshield. In fact, they appear to follow flow streamlines which do not change over time. The air velocity at any point (x, y, z) from the driver's perspective remain constant in time, thus simply by changing the observer's perspective, the same flow field can be changed from unsteady to steady.

It should be obvious that equations simplify greatly when a flow field is steady simply due to the fact that one degree of freedom (in this case, time) has been removed from the problem. Aeronautical engineers take advantage of this aspect when analyzing flow over moving objects. Instead of analyzing the unsteady flow field around an airplane, they convert the problem to a steady flow. When flying at speed V_{flight}, they imagine it flying into a head wind of magnitude V_∞ equal in magnitude and opposite in direction to V_{flight}. In that case the airplane makes no headway, thus appearing to sit still in the oncoming air. The streamline trajectory of snowflakes in such a headwind will look the same to a ground observer as to the pilot inside the airplane.

All of the flow systems used as examples so far may be referred to as Eulerian systems. In this type of flow field, the observer decides to investigate a specific region in (x, y, z) while the fluid flows through it in time, t. There is another way to look at flow fields where the observer chooses to "ride along" with a fluid packet and study exactly what happens to it as it moves along in (x, y, z, t). This method is called the Lagrangian approach. The Lagrangian approach is extremely useful in certain circumstances but may be quite difficult to calculate at times. This textbook will limit its investigations and discussions to flow field problems that are Eulerian in nature. At this point, the reader is likely to have thought about at least one stereotypical aerodynamics approach which fully utilizes Eulerian flow fields: the wind tunnel. A simplified schematic of a model in a wind tunnel test section is shown in Figure 4-1. Perhaps the most important concept to keep in mind when considering wind tunnels is the following: whether one studies the aerodynamic forces resulting from the *movement* of an object moving at speed, $|V|$, through a *stagnant* fluid or one studies aerodynamic forces on a *fixed* object with fluid *moving* over it at the same speed, $|V|$, the aerodynamic forces exerted on the object in either case will be exactly the same. Engineers always prefer to approach problems in the simplest way, and aerospace engineers are no exception.

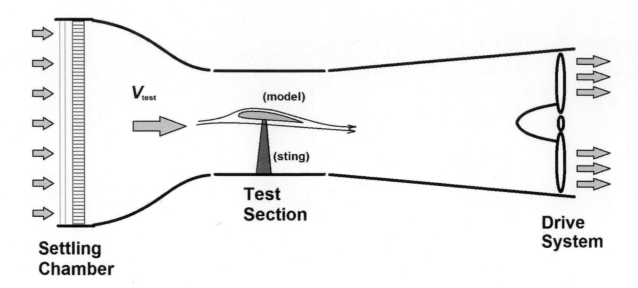

Figure 4-1: The modern wind tunnel allows aerospace engineers and experimentalists to evaluate flow fields using a steady (or even unsteady) Eulerian approach. The model to be tested is fixed relative to the observer in the test section while a fluid such as air is driven over the model.

Referring to Figure 4-1, the wing in the test section (TS) and the flow surrounding it is steady relative to the observer, making data at different stations easy to collect and recollect at different times with the same results. Now one may simulate the wing in a forward motion going at a flight velocity, V_{flight}, by simply holding it steady and moving air over it in the other direction at a test section velocity, V_{test}. The relationship describing these two velocities is simply

$$\vec{V}_{test} = -\vec{V}_{flight}$$

To physically measure the values of p, T and V, a model of the airplane is mounted inside a wind tunnel equipped with a variety of sensors. In such a tunnel a drive system fan generates the wind velocity, V_{test}. The fan can be mounted either in front or behind the model, but to minimize the effect of fan wake turbulence on the airplane, the aft position is best. Wide areas known as settling chambers are often placed forward of the TS to condition the upwind flow by running the intake air through one or more fine mesh screens (to reduce freestream turbulence) as well as a honeycomb mesh (to straighten the flow). As a result of the addition of flow conditioners and friction on the inner walls of the wind tunnel, the power required to drive the wind tunnel fan far exceeds the power needed to propel the model. The lift and drag force on the model, as well as the pitch, roll and yaw moments can be measured by suspending the airplane from an instrumented balance called a sting. Frequently those measurements are made on small-scale models of the airplane to keep the wind tunnel size down and make testing less costly. Similarly automakers test cars in wind tunnels to improve the streamlining of the body and to measure the drag of the car over its operating speed range. The presence of the ground must be simulated, as it affects the flow field over the car. In small models, the boundary layer formed by the wind blowing over the floor can adversely alter the car drag measurements. To minimize that problem, some wind tunnels replace the floor by a moving belt. Others test two car models simultaneously, each one mounted to the other base-to-base to form a mirror image of the other. In this way the stream plane in between the two cars is forced to remain parallel to the floor. Instead of using snow for flow visualization, streamlines can be made visible using smoke as shown in Figure 4-2.

Figure 4-2: Full scale model under test in 18 ft x 34 ft wind tunnel at the General Motor Technical Center in Warren, MI.

4.3   MOLECULAR COMMUNICATION AND MACH NUMBER

Before going any further in flow analysis, it is important to underscore a fundamental behavior of all gases – that of molecular communication. The presence of objects in an oncoming flow must be communicated upstream so that the molecules can "know" to get out of the way of the object and flow around it. Acoustic, pressure, and rarefaction waves transmit information at a speed which depends solely upon the mean random kinetic energy of the particles within the flow (recall this is measured by temperature, T). The speed at which all gas molecules transfer information is limited by the local speed of sound, a, which may be calculated as

$$a = \sqrt{\gamma RT} \tag{4.1}$$

The specific heat ratio, γ, compares the energy capacity of a gas if it is allowed to do work while absorbing heat to the energy capacity of the same gas in a state where it is unable to do work while absorbing heat. This will be discussed in more detail later, but for the current study, suffice to say that for air (or any other diatomic molecule at relatively normal temperature ranges)

$$\gamma_{air} = 1.4$$

Cases involving subsonic flight can become quite complex in their flow field behavior precisely due to the fact that the local region around and in front of an object moving through a fluid is overwhelmed with information regarding the whereabouts of the solid object and its motion. This upstream communication results in the fluid moving out of the way of the oncoming object. In cases of supersonic flight, the fluid is still provided a mechanism for receiving information about an oncoming object soon enough to move out of its path. The phenomenon known as a shock wave develops around supersonic aircraft, missiles, and bullets to enforce the continuity of the physical laws that govern the universe by setting up a region of incredible entropy generation to provide subsonic flow close to the object's surface.

The careful reader may be wondering whether there is any way to firmly establish a useful reference condition within a generic flow field, be it subsonic or supersonic, steady or unsteady, and so on. To avoid the trappings of using reference conditions measured near an object in motion (the local flow field near an object will undoubtedly be heavily influenced by the object's presence and thus will have fluctuating state variables), a station "infinitely far upstream" is said to exist whereby the fluid variables of the flow moving toward the object are not yet "aware" of its presence, and thus may be used as satisfactory reference values. This station is referred to as the "infinity station", and the state variables there are specially denoted by the infinity subscript. These values are often referred to as the "freestream" values indicating their freedom from influence of the object immersed within it.

Consider an airplane operating at steady flight speed V_{flight} through "still air" at a specified altitude in a Standard Atmosphere. From the simplified Standard Atmosphere tables in Chapter 3, read the values of p, T and ρ at that altitude and refer to these as "freestream" values:

a. freestream pressure at infinity, p_∞
b. freestream temperature at infinity, T_∞
c. freestream density at infinity, $\rho_\infty = p_\infty/(RT_\infty)$

The actual pressure, p, temperature, T, and density, ρ, elsewhere in the flow field around the airplane are different from those far upstream at infinity. These values are also called "static", but depend on the local molecular number density and their random velocity vector, c. These "static" values are not to be confused with "total" reference values, p_0, T_0 and ρ_0, to be identified later.

The equations to use for the theoretical analysis depend on the freestream flight Mach number (named after the famous Austrian physicist, Ernst Mach, who became the first investigator to capture photographs of shock waves around supersonic bullets), M_∞, which is the ratio of V_∞ to the free stream speed of sound, a_∞, or

$$M_\infty = \frac{V_\infty}{a_\infty} \tag{4.2}$$

To help understand how the speed of sound, the flight speed, and flight Mach number influence the flow around an object, consider a jet aircraft accelerating from rest, as shown in Figure 4-3. Initially, the aircraft is at rest, and information about the plane's presence in the local air moves out in all directions equally, creating a series of concentric rings. At half the speed of sound (i.e., $M = 0.5$), the air in front of the aircraft is still "aware" of the oncoming jet and is able to move out of the way easily. Once the plane passes through sonic and reaches supersonic speeds, the flow establishes a cone-shaped shock wave off the nose of the aircraft. In this extremely thin, highly disruptive region of flow, the air is suddenly made aware of the presence of the aircraft.

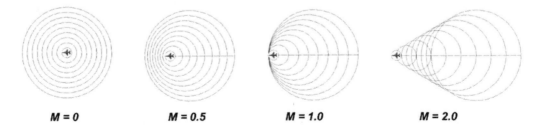

$M = 0$ $\qquad$ $M = 0.5$ $\qquad$ $M = 1.0$ $\qquad$ $M = 2.0$

Figure 4-3: As a supersonic jet fighter accelerates from rest to twice the local speed of sound, acoustic waves traveling at the speed of sound emanate into the flow field conveying information about its presence.

Although it may seem quite complicated, supersonic flow is in many ways far easier to understand in comparison to subsonic flows. The geometric simplicity of flows moving faster than the speed of sound is often a refreshing change of pace to aerodynamicists who work primarily with subsonic flows. Because supersonic flows have no way of "knowing" about the presence of an object just downstream, the flows remain absolutely straight until they reach the shock wave surrounding the oncoming object. The shock wave itself also behaves in an easy-to-predict simple geometric fashion. Consider again the motion of a supersonic jet fighter, as shown in Figure 4-4. It is clear from the simple diagram that a right triangle is formed as information travels forward and outward.

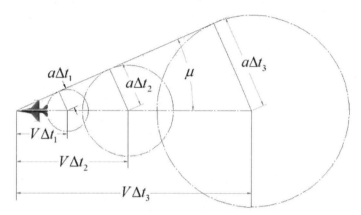

Figure 4-4: Information about the aircraft moves outward at the speed of sound, and forward at the flight velocity of the aircraft, creating a right triangle with a side of length $a\Delta t$ and a hypotenuse of length $V\Delta t$.

As shown in Figure 4-3 and Figure 4-4, the angle of the shock wave formed by nature to force the flow to adjust to the presence of an object moving faster than the speed of sound can be quickly found from the relationship of the flight speed and the local speed of sound. This angle is known as the Mach angle, μ, and is easily derived:

$$\sin \mu = \frac{a\Delta t}{V\Delta t} = \frac{a}{V} = \frac{1}{M}$$

Thus,

$$\mu_\infty = \sin^{-1}\left(\frac{1}{M_\infty}\right) \tag{4.3}$$

It is important to understand that the fundamental driver which accounts for flowing fluid is the existence of differential pressure. As the name implies, differential pressure is the spatial gradient of force per unit area due to intermolecular collisions. Put simply, pressure differential is the difference in pressure between two neighboring regions within a gas. Fluid flows move predictably along the pressure gradient, and the speed of the fluid varies directly as the rate of change of pressure per unit distance: the higher and closer the pressure difference, the faster the fluid will move. Everyone has experienced this phenomenon at some time in their lives through various everyday occurrences. For example, if the cap is suddenly released from a pressurized soda bottle, the soda will rush out from its opening and will only stop when the pressure in the bottle has equalized with the ambient air pressure. As one might anticipate, there exists a critical pressure ratio for which the velocity of the gas may be made to move sonically (i.e., at the speed of sound). For air, the ratio of the high-to-low pressure differential for sonic flow is

$$\left(\frac{p_{high}}{p_{low}}\right)_{M=1.0} = 1.893$$

For example, when a car tire is punctured, the air inside will escape at its local speed of sound, as long as the air pressure inside the tire is more than 13 psig. If the tire's gauge pressure is higher than that, the air will continue to expand to a supersonic speed, and then form a shock wave which travels faster than the speed of sound. Upon arrival at a person's ears, he or she will hear the sonic "boom" from the tire puncture. The shock wave is simply a strong compression wave which transports the pressure wave to a person's eardrums. The same phenomenon happens when an airplane flies by at supersonic speed. By the time the shock wave reaches the ear of an observer to produce the boom, the airplane has probably already flown out of sight.

Although calculations are typically much easier to make in supersonic flows, almost all objects designed to move within a flowing fluid must encounter subsonic operational conditions. For this reason, aerospace engineers must be able to predict subsonic flow behavior around most of the flight vehicles upon which they work. In some cases, subsonic flows may be simplified by making some reasonable assumptions. The major simplification which can be made in particularly low speed flows has to do with the compressibility of the working fluid. In air, for example, if the freestream flight Mach number is less than 0.3, then the change in density, ρ, is less than 5%, and the compressibility of the fluid and change in temperature is often ignored. Then ρ and T are considered known constants, and equal the freestream values ρ_∞ and T_∞.

In liquids like water, the speed of sound, a, is 1484 m/sec, at which speed the dynamic pressure, q_∞, is greater than 10,000 atm. The force required to generate these kinds of pressures is certainly excessive, thus water is not likely to ever flow near supersonic velocity and its compressibility is negligible in most cases. Likewise, air compressibility may be ignored for light airplanes, travelling at $M_\infty < 0.3$ or $V_\infty < 100$ m/sec for standard sea level air, where the speed of sound is $a_{std,sl} = 340$ m/sec. All commercial airplanes travel fast enough to experience significant changes in air density, and compressibility must therefore be taken into account in calculations. When compressibility cannot be neglected, all four independent variables (p, V, T and ρ) change significantly, thus at least four independent equations must be solved in compressible flow problems.

TEST YOUR UNDERSTANDING: ESTIMATING MACH NUMBER FROM A SONIC BOOM

A supersonic jet flies over an observer at a known altitude of 2,000 ft. Just as the aircraft passes overhead, an observer standing at sea level starts his stopwatch. He measures exactly 1.575 seconds between the jet passing directly overhead and hearing the subsequent sonic boom.

Assuming a constant speed of sound of 1,117 ft/sec, what is the flight Mach number of the plane?

Answer: $M_{flight} = 2.1$

4.4   THE CONTINUITY EQUATION (CONSERVATION OF MASS)

One of the most important concepts for engineers to master is the concept of conservation within a system. Conservation of mass within a system is especially important, because with rare exceptions, the mass within a closed system (i.e., a physical system in which mass does not cross its defining boundary) must always obey the basic principles of conservation (i.e., it can be neither created nor destroyed). This thought may be extended to include open systems (i.e., physical systems in which mass is able to freely move across the systems' defining boundaries) if one not only considers the instantaneous mass within the system, but also considers the net rate of mass moving into or out of the system. Intuitively, most people already understand this idea. For example, if marbles are dropped at a steady rate of five marbles per second into a funnel which has two hoses attached to its outlet, and three marbles per second are ejected out of one of the two hoses, then unless a buildup of marbles occurs within the funnel assembly, there must be exactly two marbles per second being ejected from the other hose.

Apply this thought now to an arbitrary volume in a fluid: imagine a stationary control volume, C$\math7{V}$, placed inside a fluid flow field as shown in Figure 4-5. It is bounded by a control surface, CS, through which fluid flows freely. As mass cannot be created or destroyed, then the time rate of change within the C$\math{V}$ must be equivalent to the difference between the rate of mass entering the C$\math{V}$ and the rate of mass leaving the C$\math{V}$, or

$$\begin{bmatrix} time\,rate\,of \\ change\,of\,mass \\ within\,C\!\!\!/\!V \end{bmatrix} = \begin{bmatrix} time\,rate\,of \\ mass\,entering \\ C\!\!\!/\!V \end{bmatrix} - \begin{bmatrix} time\,rate\,of \\ mass\,leaving \\ C\!\!\!/\!V \end{bmatrix}$$

which may be rewritten mathematically as

$$\frac{\partial}{\partial t}\iiint_{C\!\!\!/\!V} \rho\,dV\!\!\!/ = -\iint_{CS} \rho\left(\vec{V}\cdot\hat{n}\right)dA$$

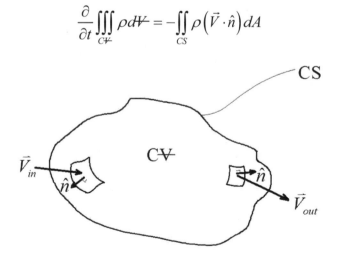

Figure 4-5: A stationary control volume within in a flowing fluid.

Consider this situation in more even mathematical detail. First, define a differential area, dA, on the CS with a unit vector, $\hat{n}$, pointing normally outward thus drawn perpendicular to area dA. Then the volumetric outflow rate through area dA is calculated as

$$\dot{V} = \left(\vec{V} \cdot \hat{n} \right) dA$$

which equals the scalar product of the velocity and unit normal vectors times the differential area. The corresponding mass outflow rate is then given by multiplying the volumetric outflow rate by the density, ρ. The mass per differential volume, dV, inside the CV is given by

$$dm = \rho \, dV$$

Integrating within the entire CV gives the mass, m, inside. Taking the partial derivative of this mass gives the time rate of mass storage inside the CV yielding the following form of the continuity equation,

$$\frac{d}{dt}(m) = \frac{\partial}{\partial t} \iiint_{CV} \rho \, dV + \iint_{CS} \rho \left(\vec{V} \cdot \hat{n} \right) dA \qquad (4.4)$$

Recall that for steady flow,

$$\frac{\partial}{\partial t}(\ ) \equiv 0$$

The surface integral can be broken into two integrals, such that

$$0 = \iint_{cs} \rho \left(\vec{V} \cdot \hat{n} \right) dA = \iint_{openings} \rho \left(\vec{V} \cdot \hat{n} \right) dA + \iint_{walls} \rho \left(\vec{V} \cdot \hat{n} \right) dA = \iint_{openings} \rho \left(\vec{V} \cdot \hat{n} \right) dA$$

Because the velocity component normal to the wall must be zero (i.e., the wall is impermeable), the surface integral over the walls is zero. The following integration is limited to steady, quasi one-dimensional flow which means that all fluid parameters (pressure, temperature, density, and velocity) are constant and uniform over each cross section and vary only in the direction of flow. If the integral is now split into an inflow opening denoted by the subscript i and an exit flow opening denoted by the subscript e, then at the inlet where the local velocity opposes the unit normal vector the dot product is negative. Likewise, at the outlet where the local velocity is in the same direction as the unit normal vector, the dot product is positive, thus,

$$\left(\vec{V}_i \cdot \hat{n} \right) < 0 \qquad\qquad \left(\vec{V}_e \cdot \hat{n} \right) > 0$$

Under these conditions, the steady flow continuity equation becomes

$$0 = -\left(\rho V dA \right)_i + \left(\rho V dA \right)_e$$

The **steady flow continuity equation then becomes a constant mass flow rate**:

$$\boxed{\rho_i V_i A_i = \rho_e V_e A_e = \dot{m}} \qquad (4.5)$$

Table 4-1 provides a brief overview of the units most often used to describe mass flow rate derived from the continuity equation in both SI and English systems of measurement.

Table 4-1: Standard units for the mass flow rate found from the continuity equation.

Variable	ρ	V	A	$\dot{m}\,(=\rho VA)$
SI Units	kg/m^3	$m/\sec$	m^2	$kg/\sec$
English Units	$slug/ft^3$	$ft/\sec$	ft^2	$slug/\sec$

It is interesting to consider the streamlines from Figure 2-6 in light of the continuity equation. The space in between each line represents a stream tube, of unit length in depth, perpendicular to the paper. Because the velocity is everywhere tangential to the walls of the stream tube, there is no fluid crossing the upper or lower walls. By conservation of mass the steady inflow rate equals the outflow rate. Assuming constant density flow, then within the stream-tube the mass flow rate is constant. One can then conclude that the local velocity V is inversely proportional to the spacing of the streamlines.

4.5 MOMENTUM EQUATION FOR INVISCID FLOW (CONSERVATION OF MOMENTUM)

Recall, Newton first proposed the now-familiar expression, F = ma. In actuality, Newton showed that

$$F = \frac{d}{dt}(mV) = V\frac{dm}{dt} + m\frac{dV}{dt}$$

Consider this statement with respect to a CV within a flowing fluid as shown in Figure 4-6. Forces act on the CV normal to the CS in the form of pressure, denoted with the subscript p, tangential to the CS in the form of shear, denoted with the subscript s. Force also acts on the mass within the CV due to gravitational acceleration, generating a body force, denoted by the subscript $body$.

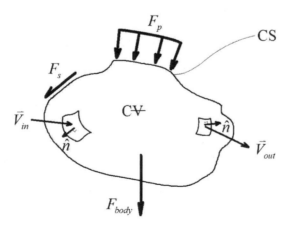

Figure 4-6: A control volume within a flowing fluid showing the various sources of forces and momentum exchange.

Just like with the continuity equation, it is helpful express the momentum balance in words:

$$\begin{bmatrix} time\,rate\,of \\ change\,of\,momentum \\ within\,C\!V \end{bmatrix} = \begin{bmatrix} time\,rate\,of \\ momentum\,entering \\ C\!V \end{bmatrix} - \begin{bmatrix} time\,rate\,of \\ momemtum\,leaving \\ C\!V \end{bmatrix} + \begin{bmatrix} net\,forces \\ acting\,on \\ C\!V \end{bmatrix}$$

Mathematically, this balance may written as

$$\frac{\partial}{\partial t}\iiint_{C\!V} \vec{V} \rho \, dV = -\iint_{CS} \vec{V} \rho\left(\vec{V} \cdot \hat{n}\right) dA + F_p + F_s + F_{body}$$

For steady flow, and in cases where one may neglect gravity (certainly justifiable for gases with small changes in overall volume height), the equation reduces to

$$F_p + F_s = \iint_{CS} \vec{V} \rho\left(\vec{V} \cdot \hat{n}\right) dA$$

Further, because pressure always acts inward on a surface and because gauge pressure is the effective pressure acting upon the surface,

$$F_p = \iint_{CS} -p\hat{n}dA = \iint_{CS} -\left(p - p_{amb}\right)\hat{n}dA$$

In the case of shear forces on the C$\!V$, it is possible to write

$$F_s = \iint_{CS} \vec{\tau} dA$$

One thing to note before proceeding is that there is a significant difference between continuity and momentum cases. *Continuity produces scalar results whereas results in momentum cases necessarily include both magnitude and direction and hence are vectors.* However, in many cases, it is possible to only consider cases of momentum balance in which the forces and flows are all aligned along a single axis, thus rendering (in effect) only scalar results. This approach is commonly referred to as the one-dimensional (1D) impulse function.

A jet engine provides a classic example of the 1D impulse function. One way to look at the jet engine is to "see" that the sum of the inward forces generated from pressure and shear inside the engine create reactions at both the inlet and exhaust (Figure 4-7). The internal workings of jet engines in practice, which has been perfected by manufacturers such as Rolls Royce, General Electric, and Pratt & Whitney, are among the most closely guarded industrial secrets of the modern age. Yet, amazingly, aerospace engineers need absolutely no information about those internal workings to determine the impulse function, and hence the thrust, of any jet engine. All that is needed is the inlet and exhaust conditions.

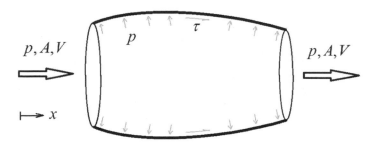

Figure 4-7: Thrust from a jet engine may be determined using a special case of the momentum equation.

Given the described jet engine, it is possible to write the impulse function as

$$\vec{T} = -\left[\iint_{CS} -\left(p - p_{amb}\right)\hat{n}dA + \iint_{CS} \vec{\tau}dA \right]_{walls} = -\left[\iint_{CS} \left(p - p_{amb}\right)\hat{n}dA + \iint_{CS} \vec{V}\rho\left(\vec{V}\cdot\hat{n}\right)dA \right]_{openings}$$

Assuming constant densities across each discreet opening, one finds the **1D impulse function**:

$$\boxed{T_x = -\left[\rho A V^2 + \left(p - p_{amb}\right)A\right]_{exit} + \left[\rho A V^2 + \left(p - p_{amb}\right)A\right]_{inlet}} \tag{4.6}$$

Generically, it is possible to write the 1D impulse function as

$$T_x = \sum I$$

where

$$I_\alpha = \dot{m}_\alpha V_\alpha + \left(p - p_{amb}\right)A_\alpha$$

For steady uniform one-dimensional flow at the openings **the impulse function vector $\vec{T}$ is always pointing inwards to the control surface regardless of the direction of the mass flow through the CS**, a fact which becomes obvious in the (albeit silly) cases of the catcher and the pitcher on skateboards shown in Figure 4-8. Note for a jet engine in flight, the exhaust impulse function is always greater than the intake impulse function, therefore the above thrust T_x is in the direction or the exhaust impulse function.

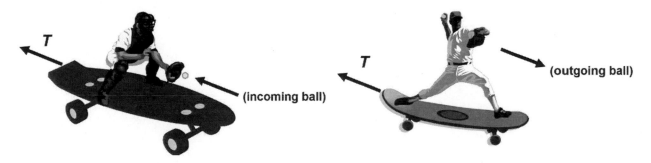

Figure 4-8: Impulse function thrust always acts in a direction pointing inward to the C¥. When the catcher on a skateboard catches a ball being thrown into the C¥, then the thrust pushes him backward, and when the pitcher on the skateboard throws the ball out of the C¥, then the thrust also pushes him backward.

4.6   THE EULER AND BERNOULLI EQUATIONS AND THE PITOT TUBE

Imagine a case where a fluid flows through a region in space with impermeable "walls" defined as that area through which the mass flow rate is identically zero. This construct is called a stream tube and a representation of such a system is shown in Figure 4-9. Now consider the behavior of the fluid between some arbitrary inlet at location 1 and outlet at location 2. The pressure at location 1 is p, and the velocity is V, while the pressure at location 2 is p plus some incremental change in pressure, dp and the velocity is V plus some incremental change in velocity, dV.

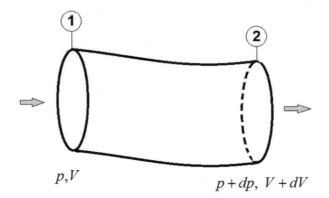

Figure 4-9: Stream tube with impermeable, inviscid walls.

Note that the flow itself is inviscid, so there is no shear stress exerted on the fluid. Next, impose upon the stream tube the additional detail that the cross sectional area normal to the fluid flow is everywhere constant within the tube. If the flow is steady, then equilibrium dictates that the sum of the forces on the system are equal to zero, or

$$\sum F_{sys} = 0$$

Using the impulse function as the basis for describing the forces on the system,

$$I_2 - I_1 = 0 \qquad \Rightarrow \qquad I_1 = I_2$$

or

$$\dot{m}_1 V_1 + \left(p_1 - p_{amb} \right) A_1 = \dot{m}_2 V_2 + \left(p_2 - p_{amb} \right) A_2$$

If the flow is steady, then no mass is building up within the tube and hence

$$\dot{m}_1 = const. = \dot{m}_2 = \dot{m}$$

With that in mind and substituting values from the diagram, find

$$\dot{m}V + \left(p - p_{amb} \right) A = \dot{m}\left(V + dV \right) + \left(p + dp - p_{amb} \right) A$$

Rearranging and simplifying terms, it is possible to arrive at the Euler equation,

$$0 = \rho V dV + dp \tag{4.7}$$

What does the Euler equation really tell engineers? Note that it reveals a fundamental relationship between density, velocity, and pressure for fluid within the stream tube. Now take this one step further and consider the case of an incompressible fluid in the stream tube. First integrate between locations 1 and 2:

$$\int_1^2 dp = -\rho \int_1^2 V dV$$

which easily integrates to give

$$\left(p_2 - p_1 \right) = \frac{\rho}{2}\left(V_1^2 - V_2^2 \right)$$

And finally arrive at a familiar expression known as the **Bernoulli equation**:

$$\boxed{p_1 + \tfrac{1}{2}\rho V_1^2 = p_2 + \tfrac{1}{2}\rho V_2^2 = const.}$$
 (4.8)

How can engineers use the Bernoulli equation? Consider a Pitot-static probe as shown in Figure 4-10:

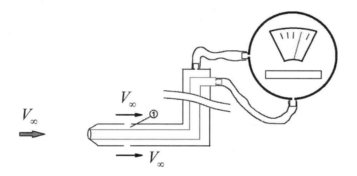

Figure 4-10: A Pitot-static probe can be used with Bernoulli's equation to indicate the velocity of a fluid.

At the tip of the probe, molecules from the flow begin to move into the central tube until they begin to pile up at one side of a diaphragm inside a pressure gauge. Once this happens, no more air molecules tend to enter the central tube and the velocity within that section goes to zero. The central tube's stagnation condition enforces that the pressure now reaches the highest value that the flow can exert, known as flow total pressure, p_0. Location 1 on the other hand still exhibits velocity at its location. The pressure within the outer shell of the probe is communicated to the other side of the pressure gauge's diaphragm so that the pressure indicated on the gauge is actually what is known as a differential pressure. This can be expressed as

$$p_{gauge} = p_0 - p_1$$

Depending upon the design of the probe, the velocity at location 1 is essentially the same as the freestream velocity value, so

$$V_1 \approx V_\infty \qquad \Rightarrow \qquad p_1 \approx p_\infty$$

Now from Bernoulli's equation, find that

$$p_1 + \tfrac{1}{2}\rho V_1^2 = p_\infty + \tfrac{1}{2}\rho V_\infty^2 = p_0 = const.$$

And since the gauge reads the pressure differential, engineers may use this fact to discern the flow's velocity (so long as they are able to appropriately estimate the density of the flow):

$$V_\infty \approx V_1 = \sqrt{\frac{2\left(p_0 - p_1\right)}{\rho}} = \sqrt{\frac{2p_{gauge}}{\rho}}$$

Pitot-static probes are commonly used on aircraft where the differential pressure gauge actually reads in terms of speed (e.g., ft/sec, knots, or mph). This is of course easily done by the manufacturer of the speed indicator, but an assumption has to be made that the density of the fluid will be that of sea level standard day air. The readout on the speed indicator is often called the calibrated airspeed, V_{cal}. However, if an airplane operates at an altitude where the true air density is below that of sea level, then the true airspeed will be greater than what is indicated, or

$$V_{true} > V_{cal} \qquad \left(\rho_{true} < \rho_{sl,std}\right)$$

This is not a problem, though, because the true airspeed may be quickly determined from the following corrective equation

$$V_{true} = V_{cal}\sqrt{\frac{\rho_{sl,std}}{\rho_{true}}} \tag{4.9}$$

4.7 INTERNAL ENERGY OF GASES

The molecules in a non-flowing gas do not sit still. They bounce around in all directions with an average velocity, $\bar{c}$, and a translational kinetic energy, $\tfrac{1}{2}\bar{c}^2$, that is proportional to its static temperature, T. The collisions that occur as a result of this motion create the fluid's static pressure, p. Molecules are simply groups of bonded atoms that come in virtually limitless forms. Each variation of molecular design allows molecules to hold energy in a manner that is directly linked to its structure. Strictly speaking, *the internal energy of gas molecules, u, is the sum of the translational, rotational, and vibrational degrees of freedom (also called "energy modes") of the molecule:*

$$u = \sum e_{mode}$$

Each mode contains an equal amount of energy per unit mass found as

$$e_{mode} = \tfrac{1}{2}RT$$

Therefore, the internal energy of a molecule may be expressed as

$$u = \left(n_{translational} + n_{rotational} + n_{vibrational}\right)e_{mode} = e_{translational} + e_{rotational} + e_{vibrational}$$

For complex molecules, vibrational modes can occur in numerous ways and make internal energy calculations more difficult. However, vibrational energy storage is typically restricted to molecules excited at very high temperatures and can thus be ignored for many aerospace engineering applications, or

$$e_{vibrational} \approx 0 \quad (\text{for "low" temperatures})$$

Neglecting vibrational modes, the internal energy per unit mass of a gas with a known chemical makeup is quite easy to determine. Consider a molecule in a simple Cartesian (orthogonal) coordinate system: since the molecule is free to move in all three (x, y, z) directions, one finds

$$e_{translational} = 3e_{mode} = \tfrac{3}{2} RT$$

The amount of rotational energy carried is a bit more complicated. For monatomic molecules (e.g., any of the noble gases or any extremely hot gases), it is not possible for the molecule to really carry noticeable rotational energy, since it is essentially spherical in representative form. Considering now a molecule with a linear structure, most often simply a diatomic molecule (i.e., a molecule made up of two atoms) in the Cartesian system, it is easy to see that the molecule may rotate with significant energy about two of its three axes (note that rotational energy along the bond-aligned axis is negligible, as shown in Figure 4-11). Finally, considering a three dimensional molecule (a nonlinear molecule with three or more atoms in its structure), all three axes may carry significant rotational energy. Summarizing the rotational energy modes,

$$e_{rotational,monatomic} \approx 0$$

$$e_{rotational,di atomic} = 2e_{mode} = \tfrac{2}{2} RT$$

$$e_{rotational,3D} = 3e_{mode} = \tfrac{3}{2} RT$$

Applying the previous equations to the three molecules shown in Figure 4-11, it is possible to quickly calculate the internal energies of the three molecules shown. Assuming each of the three molecules is not at a very high temperature, then for the helium molecule, the internal energy is simply the sum of the three translational energy modes. For the oxygen molecule (O_2), which usually appears in a diatomic form, the internal energy is the sum of the three translational energy modes plus two of the rotational modes. Finally, for the methane molecule (CH_4), the internal energy is the sum of the three translational and all three rotational energy modes.

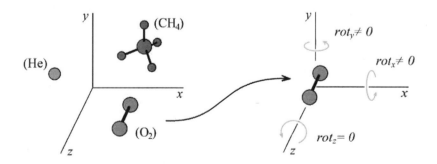

Figure 4-11: Monatomic, diatomic, and three dimensional molecules shown in the standard Cartesian coordinate system help to infer internal energy capabilities of each molecule (left). A diatomic molecule is capable of carrying rotational energy on only two axes (right).

The properties of these different molecular forms are well known for various gases, and it often helpful to simply refer to the internal energy as a function of temperature by using what is known as the coefficient of heat at constant volume, c_v, such that

$$u = c_v T \qquad (4.10)$$

Combining the concept of internal energy being carried by the sum of the different molecular degrees of freedom with the concept of coefficient of heat at constant volume relating internal energy to static temperature, one may write

$$e_{monatomic} = \tfrac{3}{2} RT \qquad \Leftrightarrow \qquad c_{v,monatomic} = \tfrac{3}{2} R$$

$$e_{diatomic} = \tfrac{5}{2} RT \qquad \Leftrightarrow \qquad c_{v,diatomic} = \tfrac{5}{2} R$$

$$e_{3D} = \tfrac{6}{2} RT \qquad \Leftrightarrow \qquad c_{v,3D} = \tfrac{6}{2} RT$$

ENGINEERING IN CODE: CALCULATING THE INTERNAL ENERGY OF A GAS

The following MATLAB® script will help the user compute internal energy for a given gas:

```
clear all;
T = input('Enter the temperature of the gas (in K):   ');
M = input('Enter the molecular weight of the gas (in kg/kmol):    ');
choice1 = menu('Molecular Model:','Monatomic','Diatomic','3D');
if choice1 == 1
    u = 1.5*(8314/M)*T;
elseif choice1 == 2
    u = 2.5*(8314/M)*T;
else
    u = 3*(8314/M)*T;
end
disp(sprintf('u = %0.0f J/kg',u));
```

Example input and resultant output (Note: "Diatomic" was selected from the pop-up menu):
```
Enter the temperature of the gas (in K):    288.16
Enter the molecular weight of the gas (in kg/kmol):    29
u = 206531 J/kg
```

4.8   HEAT COEFFICIENTS

Aerodynamics and thermodynamics rely heavily on a basic model of energy capacities of gases. The premise of the model is simple: a gases' ability to hold a certain amount of energy per unit mass directly relates to its temperature. This condition is true for any piece of matter, but unlike solids or liquids, gases take up the form of the container in which they are held and thus have the ability to exhibit significant changes in density if allowed to expand or contract. The result of this is that there are two distinct heat coefficients for gases.

COEFFICIENT OF HEAT AT CONSTANT VOLUME

The coefficient of heat at constant volume, c_v, enables engineers and scientists to evaluate the ability of a gas to absorb energy when the gas is restrained to a constant volume. A sealed rigid container of gas represents a closed constant volume gaseous system as shown in Figure 4-12. In this case, if raw energy is introduced into the gas (in the form of heat transfer) then the internal energy per unit mass, u, of the gas will increase linearly with its measured absolute temperature, T.

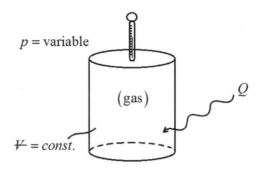

Figure 4-12: Gas in a rigid, sealed container represents a constant volume, constant mass system.

$$u = c_v T \qquad \Leftrightarrow \qquad U = mu = mc_v T = \rho \forall c_v T$$

For air,

$$c_{v,air} = \frac{5}{2} R_{air} \qquad (4.11)$$

COEFFICIENT OF HEAT AT CONSTANT PRESSURE

The coefficient of heat at constant pressure, c_p, enables engineers and scientists to evaluate the ability of a gas to absorb energy when the gas is maintained at a constant pressure. A sealed frictionless piston-cylinder assembly containing gas represents a closed constant pressure gaseous system as shown in Figure 4-13. In this case, if raw energy is introduced into the gas (again in the form of heat transfer) then the piston will rise as the gas freely expands at constant pressure and the energy per unit mass of the gas will increase linearly with its measured absolute temperature, T. This particular measure of energy per unit mass is called specific enthalpy, h.

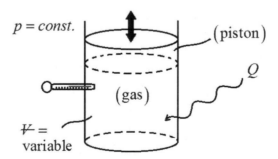

Figure 4-13: Gas in a friction-free sealed piston-cylinder assembly represents a constant pressure, constant mass system.

$$h = c_p T \qquad \Leftrightarrow \qquad H = mh = mc_p T = \rho \forall c_p T$$

For air,

$$c_{p,air} = \frac{7}{2}R \tag{4.12}$$

RATIO OF SPECIFIC HEATS

The ratio of specific heats for a gas, γ, is sometimes referred to as the isentropic expansion factor (to be discussed later in this chapter). This ratio is indicative of the amount of additional energy that may be stored in a gas if it is allowed to store some of its energy in the form of boundary work. The ratio of specific heats is simply expressed as

$$\gamma = \frac{c_p}{c_v} \tag{4.13}$$

For air,

$$\gamma_{air} = \frac{\left(\frac{7}{2}R_{air}\right)}{\left(\frac{5}{2}R_{air}\right)} = 1.4 \tag{4.14}$$

4.9   ENERGY EQUATION FOR FLOWING GAS (CONSERVATION OF ENERGY)

Just as mass and momentum within fluid systems can be shown to obey simple conservation principles, it is not difficult to extend this idea to energy in the fluid as well. Typically, fluids tend to store energy as internal, kinetic, and potential energies. Energy in these three areas may be transferred by means of transferring mass across the system boundaries. Energy may also be transferred without moving mass through fluidic work and heat transfer. The modes of energy storage and transfer may be summarized as:

- U – **Internal energy** of a gas comes from the random translational, rotational, and vibrational behavior of gas molecules. Internal energy is generally given as

$$U = mc_v T$$

- KE – **Kinetic energy** of a gas comes from two distinct sources: the kinetic energy associated with the internal energy of a gas specifically addresses the gas molecules' random motion, but if the gas itself has a prevailing direction and magnitude of motion (referred to as bulk velocity), this adds another source of kinetic energy storage to the gas. Kinetic energy is generally given as

$$KE = \tfrac{1}{2}mV^2$$

- PE – **Potential energy** is something that all gases exhibit in a gravitational field simply due to the fact that the gas must have some mass per unit volume (i.e., density) which is attracted to nearby massive bodies. Potential energy is generally given as

$$PE = mgh$$

- **W – Work energy** of a gas comes in two primary modes: flow work (work done between a fluid and its surroundings through shear and pressure), and shaft work (work done between a fluid and an often-rotating mechanical structure). Work energy is considered positive if the gas transfers energy into its surroundings (i.e., "does work on the surroundings"). Work energy is generally given as

$$W = W_{flow} + W_{shaft}$$

- **Q – Heat energy** (energy in its "raw" form) flows as a result of temperature differentials existing within a fluid system. Heat energy always moves from hot to cold. Heat energy is considered positive if the gas receives energy from its surroundings.

To derive the conservation of energy equation first construct a tube of constant cross-sectional area, A, and length, s, as shown in Figure 4-14. The tube exhibits flow of a fluid through its interior.

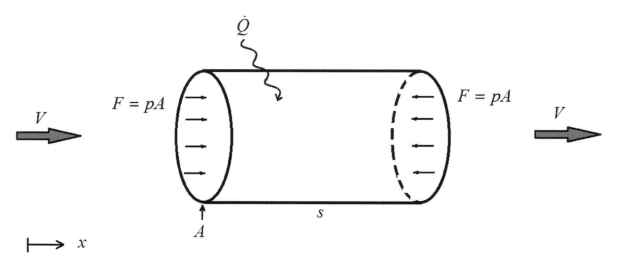

Figure 4-14: Fluid carries and transfers energy while moving through a constant-area tube.

Next express in words how the rates of energy flow may be conserved in this system (denoted by subscript "*sys*"):

$$\begin{bmatrix} time\ rate\ of \\ change\ of \\ energy \end{bmatrix}_{sys} = \begin{bmatrix} heat\ added \\ (from \\ surroundings) \\ to\ system \end{bmatrix} - \begin{bmatrix} work\ done \\ (on \\ suroundings) \\ by\ system \end{bmatrix} + \begin{bmatrix} rate\ of\ internal, \\ kinetic, and \\ potential\ energy \\ entering\ system \end{bmatrix} - \begin{bmatrix} rate\ of\ internal, \\ kinetic, and \\ potential\ energy \\ leaving\ system \end{bmatrix}$$

Mathematically, this can be stated

$$\left. \frac{dE}{dt} \right|_{sys} = \dot{Q} - \dot{W} + \sum_{in} \dot{m} \left(c_v T + \frac{V^2}{2} + gz \right) - \sum_{out} \dot{m} \left(c_v T + \frac{V^2}{2} + gz \right)$$

NOTE: **Heat in** is considered **positive** and **work out** is considered **positive**!

If the additional conditions are imposed that the system is steady, adiabatic, and that the gravitational effects are negligible, the equation may be reduced to

$$\dot{W} = \sum_{in} \dot{m}\left(c_v T + \frac{V^2}{2}\right) - \sum_{out} \dot{m}\left(c_v T + \frac{V^2}{2}\right)$$

The dot symbol above the heat and work energy terms indicates the time derivative of each, or

$$\dot{W} \equiv \frac{d}{dt}(W) \qquad \dot{Q} \equiv \frac{d}{dt}(Q)$$

Thus, $\dot{W}$ is simply the summation of all the types of work rate (i.e., power) that this system may do (namely from shaft work, flow work, and boundary work). Noting that there is no shaft in the system and that the boundary does not move, only flow work is possible in the system. How may one express flow work rate? Consider that this type of work is essentially a "force times distance" type of work, so

$$\dot{W} = \frac{d}{dt}\left(W_{flow}\right) = F\left(\frac{ds}{dt}\right) = pA\left(\frac{ds}{dt}\right) = pAV$$

The last expression for flow work rate is reminiscent of the mass flow rate equation...

$$\dot{m} = \rho AV \qquad \Rightarrow \qquad AV = \frac{\dot{m}}{\rho}$$

So

$$\dot{W}_{flow} = p\left(\frac{\dot{m}}{\rho}\right) = p\dot{m}\left(\frac{RT}{p}\right) = \dot{m}RT$$

Note that one could also determine the work rate done by the system by investigating the differential between the rate of work done by the flow at the inlet and work done by the flow at the exit, or

$$\dot{W} = \dot{W}_{flow,out} - \dot{W}_{flow,in}$$

Here, note that if $\dot{W} < 0$, it implies that $\left|\dot{W}_{flow,in}\right| > \left|\dot{W}_{flow,out}\right|$, thus the system is accepting work done on it by the surroundings. Likewise, if $\dot{W} > 0$, it implies that $\left|\dot{W}_{flow,out}\right| > \left|\dot{W}_{flow,in}\right|$, thus the system is doing work on the surroundings. Resuming the evaluation of flow work, one may write

$$\dot{W} = \dot{W}_{flow,out} - \dot{W}_{flow,in} = \dot{m}\left(RT_{out} - RT_{in}\right)$$

Substituting the work expression back into the original energy balance, find

$$\dot{m}\left(RT_{out} - RT_{in}\right) = \dot{m}\left(c_v T + \frac{V^2}{2}\right)_{in} - \dot{m}\left(c_v T + \frac{V^2}{2}\right)_{out}$$

Simplifying,

$$c_v T_{out} + R T_{out} + \frac{V_{out}^2}{2} = c_v T_{in} + R T_{in} + \frac{V_{in}^2}{2} = const.$$

Note that since

$$c_v + R = c_p$$

It is possible to rewrite the energy balance expression as

$$c_p T_{out} + \frac{V_{out}^2}{2} = c_p T_{in} + \frac{V_{in}^2}{2} = c_p T_0$$

Leading to a concise expression for the **energy equation for steady, adiabatic, gas flows**:

$$\boxed{T_0 = T_{out} + \frac{V_{out}^2}{2c_p} = T_{in} + \frac{V_{in}^2}{2c_p} = T_\alpha + \frac{V_\alpha^2}{2c_p}} \qquad (4.15)$$

Note that for air (or any other diatomic gas), it is also possible to write

$$T_0 = T_\alpha + \frac{V_\alpha^2}{7R} \qquad \left(diatomic\ gases\ only! \right)$$

Eq. (4.15) is the most frequently used format for the energy equation. The stagnation temperature, T_o, is also called "total" temperature. It is the maximum temperature in the entire flow field, which occurs everywhere the flow is stopped. At the stagnation point all directional kinetic energy is converted into molecular internal energy. If the deceleration is reversible and adiabatic, then at the stagnation point the molecular random motion velocity, c, is maximum and the pressure rises to the "total" pressure, p_o.

4.10   ISENTROPIC FLOW

In many cases, aerospace engineers opt to work with idealized conditions and flows most of the time. The ideal state of flow reversibility is sometimes assumed along with the condition that no heat is gained or lost for the system. *A flow in which no frictional losses are exhibited is known as a reversible system.* Additionally, *a system exhibiting a lack of heat exchange across its boundaries is referred to as an adiabatic system.* Furthermore, *a process that is both adiabatic and reversible is known as an isentropic system.* Consider a case in which a specified heat exchange into a closed, frictionless (reversible) piston-cylinder system (Figure 4-15) occurs:

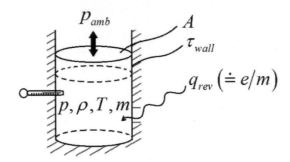

Figure 4-15: Gas inside a sealed, friction-free piston-cylinder assembly represents a closed reversible system.

Now if one is to add some incremental reversible heat, dq_{rev}, to the cylinder, then for the closed system, the energy balance would read:

$$\begin{bmatrix} internal \\ energy \\ in\ C\!\!\!V \end{bmatrix} = \begin{bmatrix} energy \\ into \\ C\!\!\!V \end{bmatrix} - \begin{bmatrix} energy \\ out\ of \\ C\!\!\!V \end{bmatrix} + \begin{bmatrix} work\ done \\ on\ C\!\!\!V\ by \\ surroundings \end{bmatrix} - \begin{bmatrix} work\ done \\ by\ C\!\!\!V\ on \\ surroundings \end{bmatrix}$$

or, mathematically speaking,

$$dE = dQ_{rev} - 0 + dW_{net}$$

Per unit mass, this can be rewritten as

$$de = dq_{rev} - 0 + dw_{net} \tag{4.16}$$

where dw_{net} represents the incremental net work done by the system per unit mass. One boundary condition which is absolutely essential is that the internal pressure, p, and the external pressure, p_{amb}, are identical. This is enforced as long as there is no friction force on the walls. Recall from the state equation that at state 1 (i.e., at the state just before the reversible heat was introduced),

$$p_1 = \rho_1 R T_1$$

Just to consider one possible outcome, assume for a moment that once the incremental heat is added to bring the contents of the cylinder to state 2, the piston does not move. Recall that the pressure has to be equal to the ambient pressure still, so

$$p_2 = \rho_2 R T_2 = p_1 = \rho_1 R T_1$$

Note that since the piston is assumed not to have moved, then

$$\rho_2 = \rho_1$$

but since

$$p_2 = p_1 = p_{amb}$$

and

$$R = const.$$

Then the temperature at states 1 and 2 must also be the same, or

$$T_2 = T_1$$

Note also that if the piston does not move, then there is no boundary or flow work, so

$$W_{net} \equiv 0$$

And now Eq. (4.16) may now be written

$$de = dq_{rev} = du = d\left(c_v T\right) = c_v dT \tag{4.17}$$

But this assumption leads to an impossible situation – one may easily recognize from Eq. (4.17) that for any incremental reversible heat addition, there *must* be some similarly signed incremental change on temperature for a system. If that is the case, then

$$T_2 \neq T_1 \qquad \Rightarrow \qquad \rho_2 \neq \rho_1$$

All this proves that the initial assumption (i.e., the assumption that the piston would not move once reversible heat was added to the system) is not physically possible. Thus, for some non-zero incremental reversible heat addition the frictionless piston *absolutely must move* and there is some resultant incremental boundary work done by the control volume on the surroundings, so

$$de = dq_{rev} - dw \qquad \Rightarrow \qquad dq_{rev} = c_v T + dw$$

Since there is piston motion, there will necessarily be flow work. Solving for the work done by the system,

$$W = Fs \qquad \Rightarrow \qquad w = \frac{W}{m} = \frac{Fs}{m} = \frac{pAs}{m}$$

noting that

$$As = \Delta V$$

91

So

$$dw = d\left(\frac{p\cancel{V}}{m}\right) = d(RT) = d(pv)$$

Now examine the right hand side (RHS) of the last equation:

$$d(pv) = pdv + vdp = RdT$$

Noting that

$$v = \frac{1}{\rho} = \frac{RT}{p}$$

One may now write

$$pdv = RdT - \left(\frac{RT}{p}\right)dp$$

so

$$dq_{rev} = c_v T + dw = c_v dT + RdT - RT\left(\frac{1}{p}\right)dp$$

$$dq_{rev} = (R + c_v)dT - RT\left(\frac{1}{p}\right)dp = c_p dT - RT\left(\frac{1}{p}\right)dp$$

Next consider what may happen if the system is adiabatic so that the incremental reversible heat into the system is now zero:

$$0 = c_p dT - RT\left(\frac{1}{p}\right)dp$$

Rearranging terms and integrating from state 1 to state 2 yields

$$\int_1^2 \frac{c_p}{RT}dT = \int_1^2 \left(\frac{1}{p}\right)dp \qquad \Rightarrow \qquad \frac{c_p}{R}\ln\left(\frac{T_2}{T_1}\right) = \ln\left(\frac{p_2}{p_1}\right)$$

which may now be expressed as

$$\left(\frac{T_2}{T_1}\right)^{\frac{c_p}{R}} = \left(\frac{p_2}{p_1}\right)$$

Examining the exponent term on the left hand side (LHS),

$$\frac{c_p}{R} = \frac{c_p}{c_p - c_v} = \frac{c_p/c_v}{c_p/c_v - c_v/c_v} = \frac{\gamma}{\gamma - 1}$$

So finally one may arrive at an expression which relates pressure and temperature at varying states in a frictionless adiabatic (i.e., isentropic) system. These relationships are often referred to as the **isentropic relations**:

$$\boxed{\left(\frac{p_2}{p_1}\right) = \left(\frac{T_2}{T_1}\right)^{\frac{\gamma}{\gamma-1}}} \tag{4.18}$$

or, written another way,

$$\boxed{\left(\frac{T_2}{T_1}\right) = \left(\frac{p_2}{p_1}\right)^{\frac{\gamma-1}{\gamma}}} \tag{4.19}$$

Note that this may also be written as

$$p_0 = p_1 \left(\frac{T_0}{T_1}\right)^{\frac{\gamma}{\gamma-1}} \tag{4.20}$$

The main idea with the previous derivation is to get to a point where one may reliably relate p, V, T (and ρ) in situations where Bernoulli's equation may run into trouble. Errors in calculations for gaseous flows may occur using Bernoulli's equation in any flow which is significantly compressible (for gases, this occurs at $M_\infty = 0.3$), where errors in density exceed 5% and create errant values in resultant pressures, temperatures, or velocities. In many introductory courses in aerospace engineering, flows are typically assumed to be isentropic. Whenever a flow can be treated thus, it is always safest to use the isentropic relations when relating temperature and pressure. Unfortunately, if one needs to relate velocity to pressure, this will mean adding an additional calculation step as compared to the one-step simplicity of pressure-to-velocity offered by Bernoulli's equation:

$$p \rightarrow V \Rightarrow \qquad p_0 = p_1 + \tfrac{1}{2}\rho V_1^2 \qquad \left[Bernoulli's\ Eq.:\ M_\infty < 0.3\ only!\right]$$

$$p \rightarrow V \Rightarrow \quad \begin{cases} p \rightarrow T \Rightarrow p_0 = p_1 \left(\dfrac{T_0}{T_1}\right)^{\frac{\gamma}{\gamma-1}} \\[2em] T \rightarrow V \Rightarrow T_0 = T_1 + \dfrac{V_1^2}{2c_p} \end{cases} \qquad \left[Isentropic + Energy\ Eq.s: any\ M_\infty\right]$$

EXAMPLE 4.1: PREDICTING THE PITOT-STATIC READING ON A SUPERSONIC JET

Given:

A McDonnell-Douglas F4 Phantom fighter aircraft takes off from a runway at sea level on a standard day. Under these conditions, the pilot must reach 201 mph for safe lift off.

Find:

a) What is the take-off Mach number? Is this speed OK to use Bernoulli's equation?
b) Using an ordinary Pitot-static probe, what should the pressure differential read before the pilot pulls back on the yoke (use isentropic relations to calculate)?
c) What would the predicted reading be using Bernoulli's equation?

Solution:

a) $\qquad M_{TO} = \dfrac{V_{TO}}{a_\infty}$

$$V_{TO} = \left(\frac{201\,mi}{hr}\right)\left(\frac{5280\,ft}{sec}\right)\left(\frac{hr}{3600\,sec}\right) = 295\,ft/sec$$

$$a_\infty = \sqrt{\gamma R T_\infty} = \left[\left(1.4\right)\left(\frac{1716\,ft\cdot lb_f}{slug\cdot °R}\right)\left(519°R\right)\left(\frac{slug\cdot ft}{lb_f\cdot sec^2}\right)\right]^{\frac{1}{2}} = 1116\,ft/sec$$

$$M_{TO} = \frac{295}{1116} = \boxed{0.264} \qquad\qquad M_{TO} < 0.3 \Rightarrow \boxed{Bernoulli's\,OK,\,but\,will\,have\,some\,error\,!}$$

b) $\qquad p_{diff} = p_0 - p_\infty; \qquad\qquad p_0 = p_\infty\left(\dfrac{T_0}{T_\infty}\right)^{\frac{\gamma}{\gamma-1}}$

$$T_0 = T_\infty + \frac{V_{TO}^2}{7R} = 519°R + \left(\frac{(295)^2\,ft^2}{sec^2}\right)\left(\frac{slug\cdot °R}{7(1716)\,ft\cdot lb_f}\right)\left(\frac{lb_f\cdot sec^2}{slug\cdot ft}\right) = 526.2°R$$

$$p_0 = 2116\,psfa\left(\frac{526.2}{519}\right)^{3.5} = 2116\,psfa(1.05) = 2221\,psfa$$

$$p_{diff} = (2221 - 2116)\,psfa = \boxed{105\,psfa}$$

c) $\qquad p_{diff} = p_0 - p_\infty = \tfrac{1}{2}\rho_\infty V_\infty^2$

$$p_{diff} = \left(\frac{1}{2}\right)\left(\frac{0.002377\,slug}{ft^3}\right)\left(\frac{(295)^2\,ft^2}{sec^2}\right)\left(\frac{lb_f\cdot sec^2}{slug\cdot ft}\right) = \boxed{103\,psfa}$$

$$\left(underpredicts\ p_{diff}\ by \sim 2\%\right)$$

Check:

The values are relatively close, seem to be appropriate in magnitude, have the correct units, and address the questions posed.

4.11 DIMENSIONAL ANALYSIS

INTRODUCTION

A major part of aerodynamics research involves the analysis of the effects of rather large groups of parameters on aerodynamic forces or similar phenomena. As one might imagine, this process can become quite complicated when there are more than just three or four varying flow field parameters. Even worse, a difficult situation can arise when applying the studied effects to a new engineering problem: how can one apply things that have been studied to a new situation if none of the specific parameters in the new flow resemble conditions that have already been explicitly studied?

The way engineers solve this apparent dilemma is by using what is known as dimensional analysis. This process involves the grouping of various terms together to form new conglomerated terms which exhibit no units. Oftentimes, these groups are simply ratios of two terms which have the same fundamental units. For a wing, for example, one can conceive of a simple comparison between lift and drag. By dividing the lift force by the drag force, one is left with a dimensionless number which provides designers with a way of establishing a type of "performance" number for the wing.

COLLECTING DIMENSIONLESS TERMS & THE BUCKINGHAM PI THEOREM

Though it may not be immediately obvious, the successful collection of relevant dimensionless terms for a given engineering scenario is sometimes a matter of experience or even luck. Fortunately, there is a fairly straightforward method which can be used to quickly determine dimensionless terms. The process can be described according to the following steps:

1. Collect the list of all (apparently) independent relevant flow parameters.
2. Express each parameter in its fundamental dimensions (*in either mLt or FLt, but not both!*).
3. Determine the minimum number of dimensions needed to express the whole set of parameters.
4. Select as many *independent* variables (called "repeating variables") as are necessary so that their combination provides a way of expressing each representative fundamental dimension.
5. Combine varying powers of the variables selected in step 4 with each of the remaining flow variables such that the result is a dimensionless term.

A natural question that may arise is, "When should I stop generating dimensionless terms?" The Buckingham Pi Theorem, which was developed in the late 1800's and formalized by Edgar Buckingham during his years at the US Bureau of Standards (now NIST), provides a way of determining the least number of dimensionless "pi terms" needed to perform dimensional analysis. The theorem states that for a given number, k, of variables (from step 1), and a given minimum number, r, of dimensions needed to express the whole set of variables (from step 3), there will need to be exactly $n = (k - r)$ dimensionless pi terms (denoted by the capital Greek letter, Π, and usually followed by a subscript number) generated (from step 5) to describe the entire system. It is important to point out that *the real "power" in dimensional analysis is that it allows engineers and scientists to evaluate behavioral relationships between variables in the most compact way*. Instead of generating huge numbers of plots, one may compress all the real trends into as little as one plot, depending upon the number of pi terms.

EXAMPLE 4.2: DIMENSIONAL ANALYSIS OF DRAG ON A SPHERE IN A FLOW

Admittedly, dimensional analysis is probably the type of thing that is best demonstrated. Consider the following example of a heavy sphere suspended by a thin wire in a wind tunnel test section as shown in Figure 4-16.

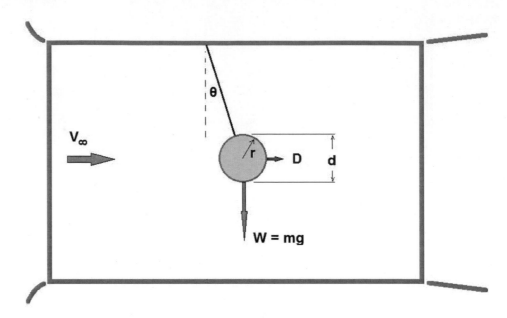

Figure 4-16: A sphere suspended by a very thin wire in a wind tunnel test section. The deflection angle, θ, may be used to deduce the aerodynamic drag exerted on the sphere.

Assuming that the diameter of the steel wire holding the sphere is extremely small in comparison to the diameter of the sphere, and due to the symmetry of the sphere, it is reasonable to conclude that the only force acting on the sphere other than its own weight is drag. Thus for a known weight and measured hang angle, drag is readily calculated as

$$D = W \tan \theta$$

Perform a dimensional analysis on this experiment:

1. <u>Collect Variables:</u> the variables that probably are important for the drag of the sphere are:
$$D = f(\rho, V, d, \mu, r) = f(\rho, V, d, \mu)$$

2. <u>Express variables in fundamental dimensions:</u>

$$D \doteq F \doteq \frac{mL}{t^2} \doteq m^1 L^1 t^{-2} \qquad \rho \doteq \frac{m}{L^3} \doteq m^1 L^{-3} \qquad V \doteq \frac{L}{t} \doteq L^1 t^{-1}$$

$$d \doteq L \doteq L^1 \qquad \mu \doteq \frac{m}{Lt} \doteq m^1 L^{-1} t^{-1} \qquad r \doteq L = L^1$$

3. <u>Minimum number of dimensions:</u> need m, L, and t to express D so minimum number is 3
4. <u>Select repeating variables:</u> for m, use ρ; for t, use V; and for L, use d

5. Combine variables for pi terms: First, note that $k = 5$ from step 1 and $r = 3$ from step 3, so the number of pi terms, $n = k - r = 5 - 3 = 2$.

Π_1 :

$$D\rho^a V^b d^c \doteq m^0 L^0 t^0 \qquad\qquad \Rightarrow \qquad \left(m^1 L^1 t^{-2}\right)\left(m^1 L^{-3}\right)^a \left(L^1 t^{-1}\right)^b \left(L^1\right)^c \doteq m^0 L^0 t^0$$

$m:$	$+1 + 1a + 0b + 0c = 0$	$\Rightarrow$	$1 + a = 0$	$a = -1$
$t:$	$-2 + 0a - 1b + 0c = 0$	$\Rightarrow$	$-2 - b = 0$	$b = -2$
$L:$	$+1 - 3a + 1b + 1c = 0$	$\Rightarrow$	$1 + 3 - 2 + c = 0$	$c = -2$

$$\Pi_1 = D\rho^{-1} V^{-2} d^{-2} = \frac{D}{\rho V^2 d^2} \qquad\qquad \leftarrow \qquad \text{(a type of "drag coefficient")}$$

Π_2 :

$$\mu\rho^a V^b d^c \doteq m^0 L^0 t^0 \qquad\qquad \Rightarrow \qquad \left(m^1 L^{-1} t^{-1}\right)^{-1} \left(m^1 L^{-3}\right)^a \left(L^1 t^{-1}\right)^b \left(L^1\right)^c \doteq m^0 L^0 t^0$$

$m:$	$-1 + 1a + 0b + 0c = 0$	$\Rightarrow$	$-1 + a = 0$	$a = 1$
$t:$	$+1 + 0a - 1b + 0c = 0$	$\Rightarrow$	$1 - b = 0$	$b = 1$
$L:$	$+1 - 3a + 1b + 1c = 0$	$\Rightarrow$	$1 - 3 + 1 + c = 0$	$c = 1$

$$\Pi_2 = \mu^{-1} \rho^1 V^1 d^1 = \frac{\rho V d}{\mu} \qquad\qquad \leftarrow \qquad \text{(known as "Reynolds number")}$$

If the data collected over a series of experimental tests was reorganized according to the two new pi terms and plotted (in this case on a log-log graph), a relationship between the pi terms would quickly emerge as shown in Figure 4-17:

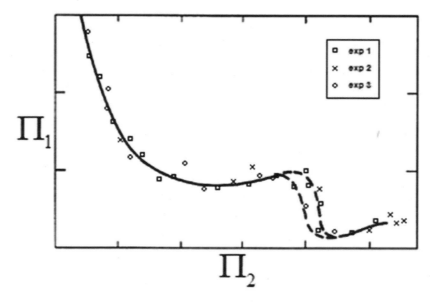

Figure 4-17: The two relevant pi terms for drag on the sphere are plotted on a graph. In this case, Π_1 is the drag coefficient, and Π_2 is the flow Reynolds number.

4.12   REYNOLDS NUMBER

Easily one of the most important dimensionless terms for a flow, particularly for the subsonic flow regime, is Reynolds number. Fundamentally, Reynolds number represents the ratio of inertial to viscous forces within a flow field. Practically speaking, Reynolds number provides a foundational metric with which to quickly predict the general local behavior of a fluid at a particular location within a flow. This dimensionless parameter owes its name to its discoverer, Anglo-Irishman (1842-1912).

Figure 4-18: Osborne Reynolds investigated the behavior of small streams of dye injected into a larger stream of flowing water.

While working on dye injection into a fluid flowing through a narrow tube, Reynolds noticed that the injected dye stream was either "direct" or "sinuous" (or transitioning between the two). He found that the critical dimensionless parameter governing this behavior was easily found by taking the ratio of the local inertial force to that of the local viscous force, such that

$$\Pi_1 = \frac{F_{inertial}}{F_{viscous}} = \frac{Aq}{\tau A} = \frac{\frac{1}{2}\rho V^2}{\mu\left(\dfrac{dV}{dx}\right)} \propto \frac{\rho V x}{\mu}$$

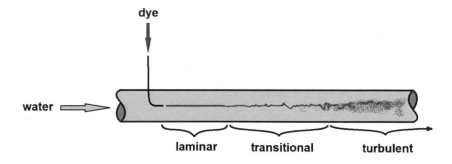

Figure 4-19: Schematic of Reynolds' dye injection experiment – Reynolds noticed distinct regions of different behavior exhibited by the injected dye stream.

For the pipe system in which Reynolds was working, he quickly discovered that the dye stream tended to remain as a cohesive structure until his dimensionless parameter had reached a critical value of about 2400. After that, the dye became intermittently wavy before finally becoming completely diffused into the water at a value of about 4000. This qualitative difference in the flow behavior seemed to have distinct quantitative values associate with it – a remarkably simple metric was suddenly discovered to anticipate general flow behavior for a given system. In later chapters, it will become evident that this behavior has an important impact on the overall influence of the fluid on bodies immersed within its flow.

All flows, whether internal as in the case of the pipe, or external as in the case of flow over a flat plate, exhibit at least laminar flow behavior. If the flow Reynolds number gets large enough, the flows will almost always transition to turbulent flow. Note that because Reynolds number is dimensionless, the values associated with laminar, transitional, and turbulent flows can be varied by modifying density, velocity, length, or fluid viscosity. This behavior is readily seen in smoke rising from a burning cigarette as shown in Figure 4-20.

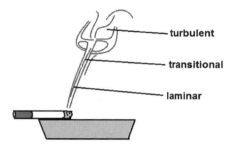

Figure 4-20: Smoke rising from a burning cigarette shows laminar and turbulent flow.

As the products of the combustion process leave the tip of the burning cigarette, several things happen:

- The flow accelerates upward, so V increases
- The flow cools down, so ρ increases
- The distance from its origination, x, increases
- Viscosity, μ, decreases slightly as $f(T)$

All these factors contribute to a dramatic rise in flow Reynolds number as the wisp of smoke travels upward and the general coherence of the smoke begins to fall apart fairly close to its source, eventually ending in a very diffuse state in the local air (anyone having worked at a job with a "smoking room" or travelled through airports with "smoking cells" will understand how diffuse smoke can become in an enclosed area).

It is critical to understand that Reynolds number can be either a local or a global value, depending upon context. It is relevant in many cases to know the Reynolds number at a specific location, but it is just as relevant in other case to know the Reynolds number based upon some global dimension (e.g., diameter of a tube or sphere, length of a flat plate or wing). The Reynolds number is generally expressed as

$$Re_x = \frac{\rho V x}{\mu}$$ (4.21)

Where x is some defining length and μ is the dynamic viscosity of the fluid. From the back page of the textbook, one may find the values for the viscosity of air at sea level in a Standard Atmosphere:

$$\mu_{Eng} = 3.7373 \times 10^{-7} \; slug/ft - \sec$$

$$\mu_{SI} = 1.7894 \times 10^{-5} \; kg/m - \sec$$

Aerospace engineers may often find themselves in a situation where it is not immediately obvious what defining length the flow Reynolds number should be based upon. In general, however, objects tend to fall into only a few broad categories – streamlined external flows, bluff body external flows, and internal flows are three of the most common categories. Consider some typical shapes and their respective lengths and reference areas shown in Figure 4-21 through Figure 4-24:

$$Re_L = \frac{\rho V_\infty L}{\mu} \qquad\qquad A_{ref} = bL = A_{side}$$

Figure 4-21: Reynolds number for external flow over a flat plate at a zero-degree angle of attack.

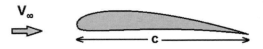

$$Re_c = \frac{\rho V_\infty c}{\mu} \qquad\qquad A_{ref} = bc = A_{planform}$$

Figure 4-22: Reynolds number for external flow over a wing section or similar streamlined body.

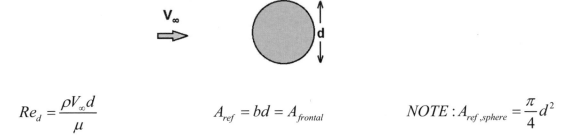

$$Re_d = \frac{\rho V_\infty d}{\mu} \qquad\qquad A_{ref} = bd = A_{frontal} \qquad\qquad NOTE: A_{ref,sphere} = \frac{\pi}{4}d^2$$

Figure 4-23: Reynolds number for external flow over an infinite cylinder or similar bluff body.

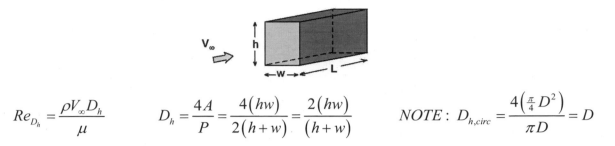

$$Re_{D_h} = \frac{\rho V_\infty D_h}{\mu} \qquad D_h = \frac{4A}{P} = \frac{4(hw)}{2(h+w)} = \frac{2(hw)}{(h+w)} \qquad NOTE: D_{h,circ} = \frac{4\left(\frac{\pi}{4}D^2\right)}{\pi D} = D$$

Figure 4-24: Reynolds number internal (pipe) flow.

One last interesting note before moving on: so far, both Reynolds number and Mach number have been shown to describe flows with some inferred characteristics. It is fascinating to note that both Osborne Reynolds and Ernst Mach were flow visualization experimentalists! They used their eyeballs as calibrated instruments to actually see valuable information about the flows they studied. All the other techniques that are employed today to solve for flows invariably call on functions governed by either (or sometimes both) these dimensionless values. All thanks to simple flow-visualization experiments.

EXAMPLE 4.3: FINDING A SPHERE'S DRAG COEFFICIENT USING REYNOLDS NUMBER

Consider the case of a smooth ball with a known geometry traveling through standard sea level air at a given speed. It is possible to use well-known drag behavior of a sphere as a function of its diameter-based Reynolds number to determine the ball's drag coefficient.

Given:

$$D = 10\,cm = 0.1\,m \qquad \rho = \rho_{std,sl} \qquad V_\infty = 100\,m/\sec \qquad \mu = 1.7894 \times 10^{-5}\,kg/m \cdot \sec$$

Find:

a) The Reynolds number of the ball
b) The approximate drag coefficient of the ball (see text, Figure 4-34)

Solution:

a)

$$Re_D = \frac{\rho V_\infty D}{\mu} = \left(\frac{1.225\,kg}{m^3}\right)\left(\frac{100\,m}{\sec}\right)\left(\frac{0.1\,m}{}\right)\left(\frac{m \cdot \sec}{1.7894 \times 10^{-5}\,kg}\right) = 6.85 \times 10^5$$

b)

from Figure 4-34,

$$C_D \approx 2 \times 10^{-1} = 0.2$$

Check:

The values seem reasonable and are appropriately dimensionless.

4.13   BOUNDARY LAYERS

In any real flow in which a viscous fluid moves over an object with some relative velocity, the particles of fluid just at the surface must exhibit no motion relative to the object's surface. This condition is known as the "no slip condition", and its impact is arguably the single most important fundamental behavior in aerodynamics and aerodynamics research. In a manner of speaking, this absolute stickiness at the surface of an object on a flow causes a chain reaction of relative speed reduction for each fluid particle, beginning with zero velocity at the surface and extending out to a location where the particle velocity eventually reaches 100% of the freestream velocity. The region of slower fluid between the surface and the location of 99% freestream velocity is known as the boundary layer (BL). Note that in some rare cases the outer edge of the boundary layer may alternately be defined as the location of 95% or even 98% freestream velocity, but 99% freestream is the most widely accepted definition of the boundary layer outer limit. A schematic view of a classic BL over a flat plate in a uniform flow field is shown in Figure 4-25.

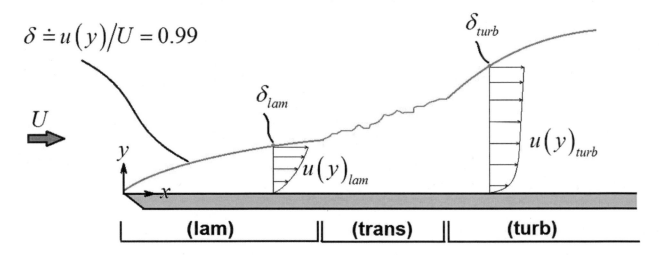

Figure 4-25: A boundary develops over the surface of a flat plate.

Why and how does this layer really form? One way of answering this question is to consider the impact of the relative motion on the fluid packet nearest to the surface. The packet exhibits a sort of torque as the lower portion of fluid refuses to move because of its proximity to the fixed surface below it and the upper portion of fluid tries to continue to move along with the rest of the fluid above it. This causes a tumbling motion called vorticity, in which the fluid packet experiences a rolling torque like a tire rolling on a surface. From Stokes' theorem, the vorticity within the BL can be shown to have a constant value. This vorticity is infinitely compact at the leading edge of the BL and tends to move upward away from the surface and into the flow as it moves downstream. The net result is a measurable growth in the BL thickness as the fluid moves downstream.

The careful aerospace engineering student may be wondering if there is some other way to picture this. Imagine a card dealer places a fresh deck of playing cards stacked neatly on a table. Now, with all of her strength, she hits the top card sideways to try and get the whole deck to move. What happens? Exactly what one would expect: only the top card, plus a few more below it tend to move and, in fact, the bottom card has not moved at all. This is a good way to visualize the no slip condition, and is presented graphically in Figure 4-26.

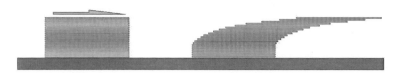

Figure 4-26: A deck of cards on a table – a push on the uppermost card demonstrates the "no slip" condition.

If that thought experiment seems really suspicious, one may rightfully point out that when a fluid encounters an object, the whole fluid parcel is moving, so just hitting the top card is really just a gimmick. Imagine another card dealer trying a similar stunt with a fresh deck of cards. This time, the dealer takes the entire deck of cards together in his hand and gently tosses the deck sideways onto the table. In this case, one would notice that the bottom card immediately "sticks" to the table, while the other cards above it slide in a diminishing relative distance between each card going from the bottom of the deck to the top. This case is shown graphically in Figure 4-27.

Figure 4-27: A deck of playing cards gently tossed onto a table help demonstrate the "no slip" condition.

Note also that in both cases of trying to move the whole deck of cards, the resultant curvature of the deck clearly shows how easy the rolling motion, or vorticity, of the fluid can get started. The cards which have been displaced can easily tumble downward given a little more forward motion. Of course, in the illustration, gravity plays a pivotal role in the physics of the cards, but it is no less a demonstration of stratified or layered friction and in some ways is quite similar to fluid motion over an object.

What is most interesting is the large difference in BL growth when investigating laminar or turbulent regions. The thickness of a laminar BL, most often referred to as δ, grows more slowly than turbulent BL thickness, and even more importantly in terms of shear stress on an object, the faster rise in the turbulent BL velocity very near the surface wall causes a much higher shear stress on the object than is experienced within the laminar BL. Quite a lot is known about the behavior in a laminar BL, but transitional and turbulent BL behavior is still a wide open field for new discoveries and modeling techniques owing to its highly random, chaotic, unsteady nature.

A few things are known about the boundary layer over a simple flat plate, which can help aerodynamicists anticipate flows in more complex surface geometries. It is known that the flow tends to stay laminar over a flat plate until the local Reynolds number reaches a value of about 100,000. As the Reynolds number increases, the effect of the inertial forces begins to completely overcome viscous forces and the BL flow begins to transition to turbulent flow. Fully turbulent flow usually always occurs on a flat plate by the time the local Reynolds number reaches a value of about 1,000,000. Thus, for a flat plate or even wings, one may use the following rule of thumb

$$Re_x < 100,000 \qquad \Rightarrow \qquad \text{laminar BL}$$
$$100,000 < Re_x < 1,000,000 \qquad \Rightarrow \qquad \text{transitional BL}$$
$$Re_x > 1,000,000 \qquad \Rightarrow \qquad \text{turbulent BL}$$

For a flat plate, the boundary layer thickness at some location, x, can be found for laminar and turbulent boundary layers as

$$\delta_{lam} = \frac{5.2x}{\sqrt{Re_x}} \tag{4.22}$$

$$\delta_{turb} = \frac{0.37x}{Re_x^{0.2}} \tag{4.23}$$

A real fluid moving relative to any surface will tend to exhibit a force resisting the relative motion as the fluid within the BL exerts shear stress on the surface of the object. Integrating the resistive stress over the entire surface area of an object produces a skin friction which contributes to the overall aerodynamic drag of the object. A large number of fluids demonstrate a linear relationship between the shear stress they exert on an object and the rate of change of velocity normal to the surface, known as the fluid strain rate. The constant of proportionality between these aspects is the fluid viscosity. Fluids that meet these criteria are known collectively as Newtonian fluids, and obey

$$\tau = \mu \frac{du}{dy} \tag{4.24}$$

Non-Newtonian fluids (i.e., those fluids whose shear stresses do not behave linearly with their strain rates) include fluids such as blood, latex paint, cheese, and even magma.

For a flat plate, the skin friction varies depending on the local boundary layer behavior and can be found as

$$C_{f,lam} = \frac{0.664}{\sqrt{Re_L}} \tag{4.25}$$

$$C_{f,turb} = \frac{0.0583}{Re_x^{0.2}} \tag{4.26}$$

Careful inspection of the schematic view of laminar and turbulent boundary layer profiles presented in Figure 4-25, along with a critical evaluation of Eq. (4.24), raises an important concept associated with shear stress of Newtonian fluids in different boundary layers: because laminar wall strain rates are less than those in turbulent flow,

$$\left(\frac{du}{dy}\right)_{lam} < \left(\frac{du}{dy}\right)_{turb}$$

It should therefore be no surprise that the skin friction coefficient will suddenly and dramatically increase once the local flow behavior has transitioned from laminar to turbulent flow as evidenced in Figure 4-28.

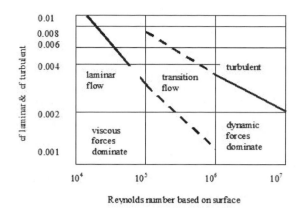

Figure 4-28: Skin friction on a flat plate as a function of Reynolds number.

It is extremely important to remember that the skin friction coefficient expresses the drag due to shear only on one side of a flat plate, so to estimate the total drag due to shear of a very thin flat plate with both sides exposed to the flow (where the entire surface is considered "wetted"), one must double the skin friction. Thus the two dimensional drag coefficient (drag coefficient per unit span) of a flat plate can be expressed as

$$C_{d,\,flat\,plate} = 2C_f$$

EXAMPLE 4.4: FINDING THE DRAG COEFFICIENT OF A FLAT PLATE IN A FLOW

Consider the case of a flat plate moving in air at a very slow speed.

Given:

$$\rho = 1.000\,kg/m^3 \qquad V_\infty = 1\,m/\sec \qquad \mu = \mu_{air} \qquad L = 1\,m$$

Find:

a) The operating Reynolds number of the flat plate
b) The overall 2D drag coefficient of the flat plate

Solution:

a)

$$Re_L = \frac{\rho V_\infty L}{\mu} = \left(\frac{1.000\,kg}{m^3}\right)\left(\frac{1\,m}{\sec}\right)\left(\frac{1\,m}{1}\right)\left(\frac{m\cdot\sec}{1.7894\times10^{-5}\,kg}\right) = 5.59\times10^4 \qquad \Rightarrow \qquad \text{laminar BL}$$

b)

$$C_d = 2C_f = 2\left(\frac{0.664}{\sqrt{Re_L}}\right) = \frac{1.328}{\sqrt{5.59\times10^4}} = 0.00562$$

Check:

The values seem reasonable and are appropriately dimensionless.

4.14   DRAG ON BODIES IMMERSED IN A FLOWING FLUID

It is important to understand that in any real flow, the interaction between the fluid and the object moving relative to the fluid will always create a force resisting the direction of relative motion. This force is fundamentally the result of billions upon billions of collisions of fluid molecules with the object's surface. As previously discussed, the normal and tangential components of force from these impacts per unit surface area are known as pressure and shear stress, respectively.

Fluid in motion over a stationary surface produces an "aerodynamic force per unit area" onto the surface. This force can be decomposed in two components:

1. *Perpendicular* to and towards the surface, where *p* is called *"static pressure"* ($-\hat{n}p$)
2. *Tangential* to and in the direction of the velocity called *"shear stress"* ($\bar{\tau}$)

Both pressure and shear stress have the same units of force per unit area.

The gas molecules in random molecular motion cause the wall shear stress. Momentum transfer during collisions with the surface causes this. For a gas at rest these collisions are specular (symmetric) and therefore produce no net force tangential to the surface, and thus no shear force. Even for a gas in motion, the magnitude of the perpendicular pressure force is many orders of magnitude greater than the shear force. A thorough understanding of the relationship between velocity and shear stress came much later than that between velocity and pressure. The following historical sequence of events is extracted from the book on "Introduction to Flight", 4th Ed. by John D. Anderson Jr.:

> *"Isaac Newton, born in England in 1642, developed the theory of universal gravitation and differential calculus. He devoted the second book of his 'Principia' 1687 to fluid dynamics-especially to the formulation of 'laws of resistance' (drag). Newton states: 'The resistance arising from want of lubricity in the parts of a fluid is proportional to the velocity with which the parts of the fluid are separated from each other'. This applies to laminar flow in now-called 'Newtonian Fluids'. Henry Pitot was born in Paris, France in 1695, he developed the Pitot tube in 1732 for the purpose to measure the water velocity in the river Seine. His calculation was based on his intuition rather than theory. The term 'Hydrodynamics' and the 'Bernoulli Equation' which provides a relationship between pressure and velocity, were first introduced by Daniel Bernoulli in his famous book 'Hydrodynamica', completed in 1734, but not published until 1738. Bernoulli was born in the Netherlands in the year 1700. In 1743 his father Johann published a book titled 'Hydraulica', where he demonstrated to have a more fundamental understanding of the relationship between fluid velocity and pressure than his son. Leonhart Euler was born in 1707 in Basel Switzerland. He worked closely together with his friend Daniel Bernoulli at the University of Basel. Euler derived the differential equation relating pressure and velocity, called the Euler Equation. This differential equation with appropriate boundary conditions allowed Jean le Rond d'Alembert to calculate the fluid dynamic forces on a body in flowing fluid, which he published in 1744. In subsequent papers he showed frustration by his inability to demonstrate that there is a drag force on a body in a flowing fluid. His 'zero drag solution' is called d'Alembert's paradox. d'Alembert was the first to formulate the wave equation in classical physics, to express the concept of a partial differential equation, the first to solve a partial differential equation by separation of variables, and the first to express the differential equations of fluid dynamics in terms of a field. He is considered one of the great mathematicians and physicist of the eighteenth century. The classical equations of motion correctly*

incorporating the effects of fluid friction due to viscosity were introduced by M. Navier, born in 1785 and further refined by Sir George Stokes, who was born in 1819. These non-linear differential equations are called the Navier-Stokes equations and are extremely difficult to solve. Only modern high-speed digital computers are able to provide solutions for 'general flow fields', and those only in terms of empirical turbulence models. Osborne Reynolds was born in 1842 in Belfast Ireland. He graduated from Cambridge University and became chair and professor at Manchester University. There he developed the Reynolds Analogy in 1874, a relation between heat transfer and frictional shear stress in a fluid. He also developed turbines and pumps and studied the propagation of sound in fluids. His most important work was published in 1883 where he introduced a dimensionless parameter (now called Reynolds Number), which determines the transition from laminar to turbulent flow. He also derived the 'Law of Flow Resistance in Parallel Channels'."

Most aerodynamic analysis is currently performed after separating the flow field of interest into two separate regions. One is a thin layer around the surface to be analyzed, which is called the boundary layer. Within it, viscous effects are dominant, but the change in pressure forces, in the direction perpendicular to the surface can be neglected. The other one is located outside the boundary layer in which viscous effects can be ignored. To find the pressure and velocity distribution surrounding the boundary layer and thus acting on the surface, use:

1. For incompressible flow – the continuity and the Bernoulli equations
2. For compressible flow – the equation of state, the continuity, energy, and either Euler or isentropic equations.

The aerodynamic force acting on a body in flight is decomposed into two components. The lift force, L, acts perpendicularly to the direction of the freestream velocity, V_∞. The drag force, D, acts in the direction of V_∞. Almost all aerodynamics research and development focus on increasing aerodynamic lift while simultaneously reducing aerodynamic drag, thus one of the main goals of aerodynamicists is to utilize shapes that exhibit high lift-to-drag (L/D) ratios.

Drag force is a combination of three main forces: parasitic drag, D_{para}, lift-induced drag (often shortened to induced drag), D_i, and wave drag, D_{wave}. Lift-induced drag is associated with finite wings and will be discussed in much more detail in Chapter 5. Wave drag is a phenomenon related to the tremendous entropy generation caused by shockwaves present in supersonic flows over bodies, the details of which are beyond the scope of this text. Parasitic drag comes from three separate sources: pressure drag, $D_{pressure}$, is due to the difference in pressure between the forward and aft-oriented faces of the object in the flow; skin friction drag, $D_{skin\,friction}$, is due to the viscous shear acting on the object within the thin region of the boundary layer; interference drag, $D_{interference}$, is due to the combination of various object geometries within a flow. Pressure and skin friction drag are often lumped together and referred to as form drag (also sometimes called profile drag), D_{form}. Pressure drag is lowest for streamlined bodies such as an airplane wing at a low angle of attack, and is highest for blunt bodies such as a truck or a bus where the wake created is even larger than its cross-section perpendicular to V_∞. Skin friction drag is lowest when the total surface area of the object exposed to the flowing fluid (known as "wetted area") is at its minimum value. Streamlining can reduce the drag of a body by a factor of ten or more. Automobiles often truncate the aft end to minimize their length for a desired volume. Eq. (4.27), Eq. (4.28), and Figure 4-29 summarize the sources of drag.

$$D = D_{para} + D_i + D_{wave} \tag{4.27}$$

$$D_{para} = D_{pressure} + D_{skin\,friction} + D_{interference} \qquad (4.28)$$

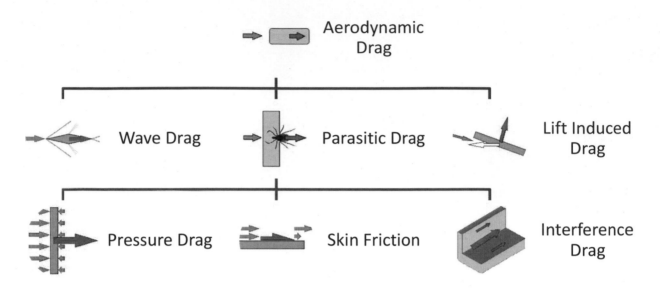

Figure 4-29: Aerodynamic drag comes from numerous sources, including shockwaves, pressure, skin friction, body interference, and lift induced drag.

Whenever an object exhibits motion relative to the fluid in which it is immersed, it will necessarily have resistance to relative motion in the form of non-zero shear and pressure forces. For 2D objects, the drag per unit span, b, may be written as

$$D' = \left(F_{p,x} + F_{\tau,x}\right)/b$$

The fact that many aerodynamic bodies have similar shapes allows engineers and scientists to test the shapes and relate their dimensionless drag with operating Reynolds number. 2D objects, though not physically possible to test, can be mimicked in a wind tunnel by extending the profile of the shape from one side of the tunnel to the other. Pressure taps and downstream measurement devices known as wake rakes can be used to ascertain the profile drag of these shapes. 3D objects are typically tested using simple force balances. In either case, the dimensionless forces are represented as drag coefficients.

$$\left(2D\ drag\right) \qquad C_d = \frac{D'}{\frac{1}{2}\rho_\infty V_\infty^2 x_{ref}} = \frac{D'}{q_\infty x_{ref}} \qquad (4.29)$$

$$\left(3D\ drag\right) \qquad C_D = \frac{D}{\frac{1}{2}\rho_\infty V_\infty^2 A_{ref}} = \frac{D}{q_\infty A_{ref}} \qquad (4.30)$$

Consider a case in which the fluid is essentially free of any viscous interaction with a surface. Imagine a circular rod of infinite span immersed in a uniform flow, as shown in Figure 4-30. The fluid slows down directly in front of the cylinder, and then accelerates around the cylinder surface. As the fluid passes the top and bottom of the cylinder, the flow reaches its maximum speed. It then decelerates as it approaches the back of the cylinder surface. In this case, the local pressure over the cylinder's surface generates opposing forces of equal magnitude front-to-back and top-to-bottom, thus the net force on the cylinder is zero.

Streamline Plot:

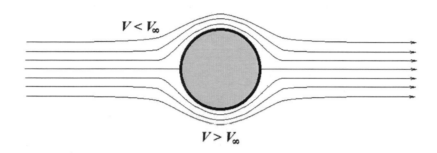

Pressure Plot:

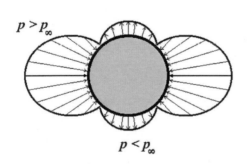

Force Plot:

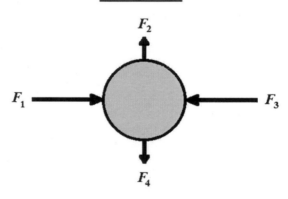

$$D = \sum F_x = F_1 - F_3 = 0 \qquad \leftarrow \qquad \text{"d'Alembert's paradox"}$$
$$L = \sum F_y = F_2 - F_4 = 0$$

Figure 4-30: Inviscid flow over a cylinder of infinite span.

Now examine a cylinder in a real flow which exerts shear stress as well as pressure on the surface of the cylinder, as shown in Figure 4-31. In this case, the fluid must lose momentum as it moves against the surface of the cylinder. By the time the fluid passes the point of maximum velocity and enters an increasing pressure gradient (i.e., pressure is increasing as the fluid particles move downstream) on the back end of the cylinder, the fluid separates away from the surface and causes a large region of swirling eddies in a pocket behind the cylinder. The pressure within this region, known as the "wake", is essentially equivalent to the ambient freestream pressure. The overall pressure and shear behavior acting on a cylinder in a viscous flow thus creates a non-zero force acting in the direction of the fluid flow. The aerodynamic force aligned with freestream velocity in any real flow is always non-zero and is known as drag.

Streamline Plot:

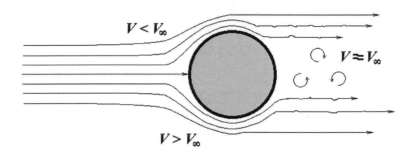

Pressure/Shear Plot:

Force Plot:

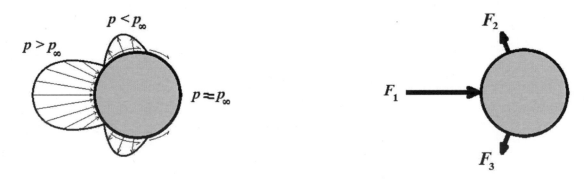

$$D = \sum F_x = F_1 - F_{2,x} - F_{3,x} \neq 0$$
$$L = \sum F_y = F_{2,y} - F_{3,y} = 0$$

Figure 4-31: Viscous flow over a cylinder of infinite span.

Consider one more interesting case in a real viscous flow in which the infinite span cylinder is rotated at a constant angular velocity in a backspin, as shown in Figure 4-32. The cylinder wall entrains the fluid near its surface and causes the fluid near the cylinder to move in a preferred direction consistent with the direction of rotation. This general flow field results in even faster fluid velocity at the top of the cylinder and slower velocity at the bottom. The pressure and shear cause a resultant force unaligned to the direction of the freestream fluid velocity. Breaking the resultant force into one component aligned with the freestream velocity and the other component normal to the freestream velocity, one finds that spinning an infinite circular cylinder in a uniform flow generates both drag and lift.

Streamline Plot:

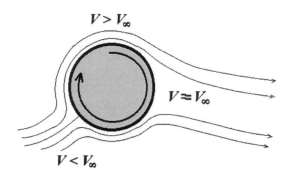

Pressure/Shear Plot:

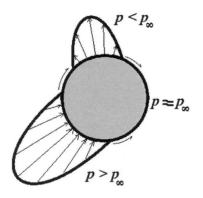

Force Plot:

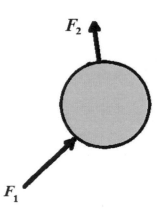

$$D = \sum F_x = F_{1,x} - F_{2,x} \neq 0$$
$$L = \sum F_y = F_{1,y} + F_{2,y} \neq 0$$

Figure 4-32: Viscous flow over a back spinning cylinder of infinite span.

The pioneering experimental work of engineer Ludwig Prandtl still provides extremely useful drag coefficient information for common 2D and 3D forms as shown in Figure 4-33 and Figure 4-34.

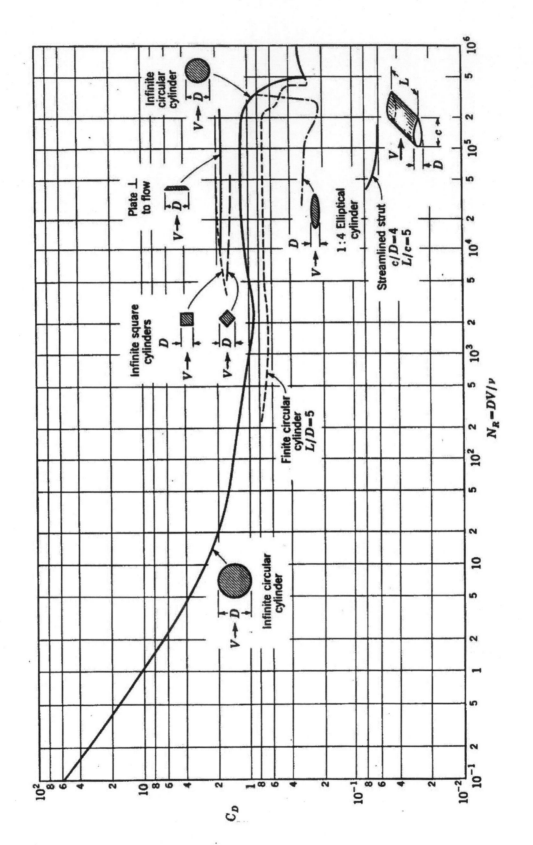

Figure 4-33: Experimental drag coefficients for two-dimensional bodies [L. Prantl, "Ergebnisse der Aerodynamischen Versuganstalt"].

112

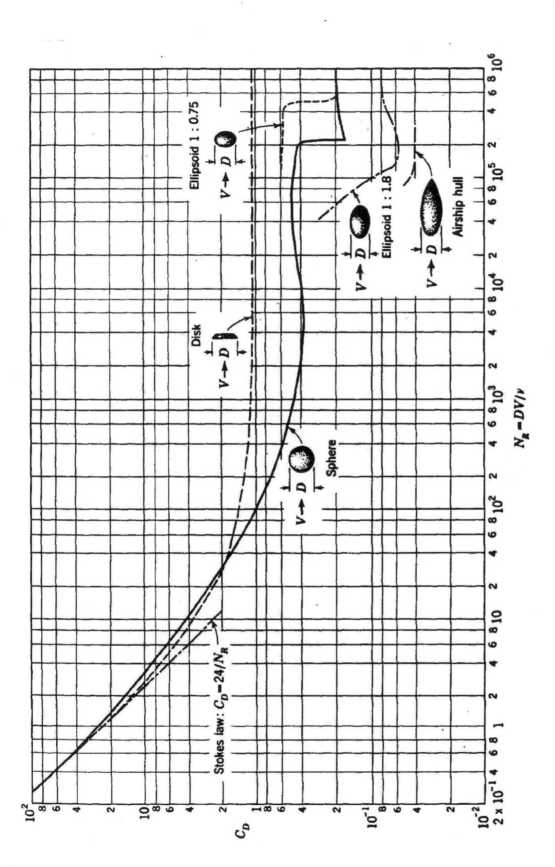

$N_R = DV/\nu$

Figure 4-34: Experimental drag coefficients of axisymmetric bodies (L. Prantdl, "Ergebnisse der Aerodynamischen Versuganstalt").

EXAMPLE 4.5: FINDING THE DRAG EXERTED ON A BALL

Given:

A paintball is fired from a pneumatic paintball marker in standard sea level air.

$$\rho = 0.002377 \, slug/ft^3 \qquad\qquad V_\infty = 290 \, ft/sec \qquad \mu = \mu_{air} \qquad d = 0.68 \, inches$$

Find:

a) The drag coefficient of the paintball
b) The drag on the paintball

Solution:

a)

first need to find Re_d :

$$Re_d = \frac{\rho V_\infty d}{\mu} = \left(\frac{0.002377 \, slug}{ft^3}\right)\left(\frac{290 \, ft}{sec}\right)\left(\frac{0.68 \, in}{}\right)\left(\frac{1 \, ft}{12 \, in}\right)\left(\frac{ft \cdot sec}{3.7373 \times 10^{-7} \, slug}\right) = 1.05 \times 10^5$$

From Figure 4-34, $C_D \approx 4.7 \times 10^{-1} = \boxed{0.47}$

b)

$$D = C_D q_\infty A_{ref} = C_D \left(\tfrac{1}{2}\rho_\infty V_\infty^2\right)\left(\tfrac{\pi}{4}d^2\right)$$

$$\tfrac{1}{2}\rho_\infty V_\infty^2 = \frac{1}{2}\left(\frac{0.002377 \, slug}{ft^3}\right)\left(\frac{(290)^2 \, ft^2}{sec^2}\right)\left(\frac{lb_f \cdot sec^2}{slug \cdot ft}\right) = 100 \, psfa$$

$$\tfrac{\pi}{4}d^2 = \frac{\pi}{4}\left(\frac{(0.68)^2 \, in^2}{}\right)\left(\frac{ft^2}{144 \, in^2}\right) = 0.0025 \, ft^2$$

$$D = (0.47)\left(100 \, lb_f/ft^2\right)\left(0.0025 \, ft^2\right) = \boxed{0.119 \, lb_f}$$

Check:

The values calculated seem to be within the appropriate range, and the answers have the right units.

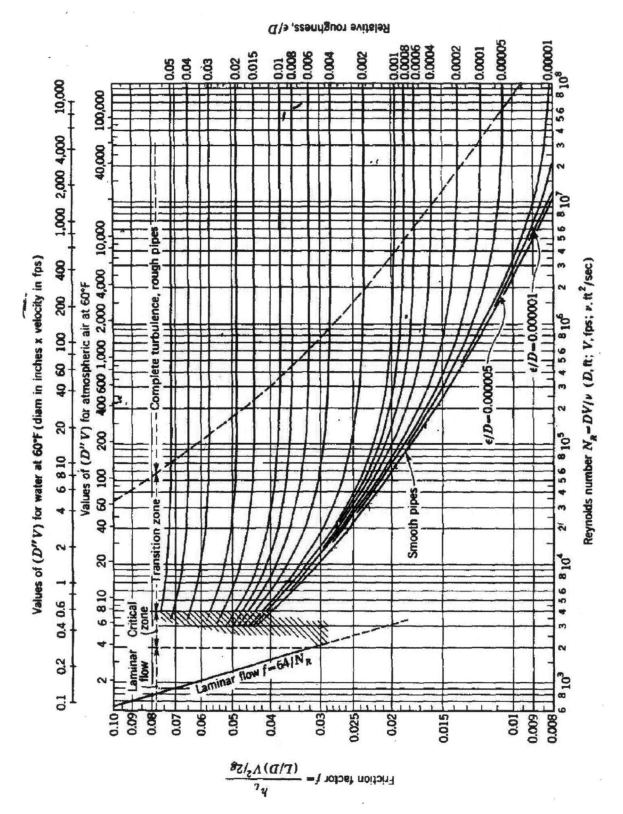

Figure 4-35: The Moody diagram for fluid flow through a pipe.

4.15   SUMMARY

STEADY AND UNSTEADY FLOWS

Unsteady flows are much harder to investigate, thus steady flows are the desirable condition.

$$\text{"steady"} \quad \Rightarrow \quad \frac{\partial}{\partial t}(\ ) \equiv 0 \qquad \text{"unsteady"} \quad \Rightarrow \quad \frac{\partial}{\partial t}(\ ) \neq 0$$

MOLECULAR COMMUNICATION

Molecules within a flow field communicate information about the state variables of the flow at the speed of sound, a, which is a local variable defined as

$$a = \sqrt{\gamma RT}$$

Supersonic flows will develop shock waves around objects immersed within the flow. The Mach number and corresponding shock wave Mach angle are given as

$$M = \frac{V}{a}; \qquad \mu = \sin^{-1}\left(\frac{1}{M}\right)$$

CONTINUITY EQUATION – MASS CONSERVATION PROBLEMS

Mass conservation problems are most often associated with things like ducted air vents or plumbing with some fluid running through them. For a steady system, whatever enters the system must also leave the system at the same rate, or

$$\sum_{in} \dot{m} = \sum_{out} \dot{m}$$

Also, for some station 1, the mass flow rate may be quantified as

$$\dot{m}_1 = \left(\rho A V\right)_1$$

MOMENTUM EQUATION – IMPULSE FUNCTION PROBLEMS

For a steady system, the change in fluid momentum will result in a non-zero net force reaction. This thrust force is the summation of the steady impulses at the entrances and exits, so that

$$T_x = \sum_{i=1}^{n} I_{x,i}$$

where the local impulse, I, at some location 1 may be calculated as

$$I_1 = \dot{m}_1 V_1 + \left(p_1 - p_{amb}\right) A_1$$

116

BERNOULLI EQUATION

The full version of the Bernoulli equation allows engineers to determine things like the pressure indicated by a manometer or the indicated airspeed of a Pitot-static probe. In any case, the complete expression is

$$p_0 = p_1 + \tfrac{1}{2}\rho_1 V_1^2 + \rho_1 g h_1 = const.$$

If the head pressure (due to gravity) is negligible in comparison to the dynamic pressure (due to velocity), the familiar expression for Pitot-static probes emerges:

$$p_0 = p_1 + \tfrac{1}{2}\rho_1 V_1^2 \qquad \rightarrow \qquad V_1 = \sqrt{\frac{2(p_0 - p_1)}{\rho_1}} = \sqrt{\frac{2p_g}{\rho_1}}$$

If gravitational effects dominate, the familiar expression for manometers emerges:

$$p_0 = p_1 + \rho_1 g h_1$$

ENERGY EQUATION

The energy equation offers a way to relate local flow field temperature to local velocity. Conservation of energy relies on the assumption that there is no heat addition to the flow under investigation. For any flow like this, there is such a thing as total temperature,

$$T_0 = T_1 + \frac{V_1^2}{2c_p}$$

For gases comprised of diatomic molecules (such as air), one may also write

$$T_0 = T_1 + \frac{V_1^2}{7R}$$

ISENTROPIC RELATIONS

The isentropic relations allow engineers to relate pressure and temperature (and also density) in an insulated shock-free flow field. The isentropic relations rely on two underlying assumptions:

1. The flow field is adiabatic (no heat is lost or gained)
2. The flow field is reversible (it is friction free and shock free)

These relations may be expressed between two generic locations 1 and 2 as

$$\left(\frac{p_1}{p_2}\right) = \left(\frac{T_1}{T_2}\right)^{\frac{\gamma}{\gamma-1}}$$

DIMENSIONAL ANALYSIS

Remember that the big deal for dimensional analysis is that it gives engineers a way to compress a large number of variables into incredibly compact dimensionless combinations for the purpose of establishing behavioral trends between these dimensionless groups. Similitude provides a way of scaling test results to predict future design performance. The two most important dimensionless parameters for an aerospace engineer are usually Reynolds and Mach numbers,

$$\text{Re}_x = \frac{\rho V x}{\mu}; \qquad M_1 = \frac{V_1}{a_1} = \frac{V_1}{\sqrt{\gamma R T_1}}$$

BOUNDARY LAYERS

Boundary layers are a necessary component of relative motion between a fluid and an object within which it is immersed. The boundary layer is often incredibly thin, and the location of its outer edge, δ, is usually defined as the location where

$$u(\delta) = 0.99 U_\infty$$

For flat plates, the boundary layer thickness growth is known to be a function of Reynolds number such that

$$\delta_{lam} = \frac{5.2x}{\sqrt{\text{Re}_x}} \qquad \qquad \delta_{turb} = \frac{0.37x}{\text{Re}_x^{0.2}}$$

The shear stress within the boundary layer acts on the surface to create skin friction drag with coefficients written as

$$c_{f,lam} = \frac{0.664}{\sqrt{\text{Re}_x}} \qquad \qquad c_{f,turb} = \frac{0.0583}{\text{Re}_x^{0.2}}$$

Remember that the skin friction coefficient equations for a flat plate are only for a single side of the plate being exposed to the fluid ("wetted"), so if both sides of a plate are exposed and one wants to find the entire drag coefficient of the plate, simply multiply the skin friction coefficient by two:

$$C_{D,\,flat\,plate} = 2C_f$$

DRAG ON BODIES IN A FLOW

Drag coefficients are most often used along with operating parameters to determine drag on an object. The drag coefficient is given as

$$(2D\,case)\ \ C_d = \frac{D'}{q_\infty l_{ref}} \qquad (3D\,case)\ \ C_D = \frac{D}{q_\infty A_{ref}} \ \rightarrow\ D = C_D q_\infty A_{ref}$$

where

$$q_\infty = \tfrac{1}{2}\rho_\infty V_\infty^2$$

CHAPTER 4 PROBLEMS

4.1) Explain why an aerospace engineer uses the velocity vector V_∞ instead of the equal and opposite flight velocity vector when studying air flow around an aircraft.

4.2) List the four variables which define an ideal friction free (Newtonian) fluid velocity flow field and provide the names and expressions of the four equations used to solve them.

4.3) A jet engine is flying through standard sea level air at $V_\infty = 140$ m/s. Its intake captures air at the flight speed in a stream-tube of area $A_\infty = 0.4 \text{m}^2$. Inside the engine fuel is added at a rate of 2.2% of the intake air mass flow rate. The burned mixture exhausts through the nozzle at $V_e = 750$ m/sec exiting at $p_e = p_\infty$ and temperature $T_e = 875$ K.

    a. Calculate the air mass flow rate entering, and the impulse function I_{in} acting on the intake (called ram drag).
    b. Calculate exhaust nozzle flow rate and impulse function I_e (called nozzle thrust).
    c. Calculate the exhaust nozzle flow density, ρ_e, exit area, A_e, and engine thrust, $T = I_e\text{-}I_{in}$

4.4) A water bottle rocket contains part water and part air which is pumped up to $p_1 = 0.827$ MPa, at negligible velocity inside. Assume the ambient air pressure is $p_\infty = 100,000$ Pa.

    a. When the water rushes out of the bottle into the ambient pressure p_∞, what will the water velocity, V_e, be in m/sec according to the Bernoulli equation?
    b. If the nozzle exit area is $A = 3 \times 10^{-5} \text{m}^2$, then what is the water exit mass flow rate?
    c. What is the rocket thrust? (*Hint:* The thrust will be equal to the impulse function at the nozzle exit, I_e)

4.5) Air is blowing out of a fan with an exit diameter, $D_e = 0.75$ m, uniform exit velocity, $V_e = 75$ m/sec, static pressure, $p_e = 100,000$ Pa and static temperature, $T_e = 30°C$.

    a. Calculate the exit air flow density
    b. Calculate the exit air mass flow rate
    c. Calculate the exit air total temperature T_o in K and total pressure p_o in Pa.

4.6) The space shuttle operates at $V_\infty = 1800$ ft/s at a high altitude where the ambient temperature is $T_\infty = 400$ °R and the ambient pressure is $p_\infty = 440$ psfa

    a. What is the flight Mach number?
    b. What is the reference total temperature $T_{o\infty}$ and total pressure $p_{o\infty}$?
    c. What is the stagnation temperature on the nose of the shuttle where $V = 0$?
    d. What will be the air temperature at point 1 on the shuttle surface where the velocity $V_1 = 1500$ ft/sec?

4.7) A high-speed airplane operates in standard sea level air at V_∞ = 960 ft/sec and has an average wing chord length c = 6 ft.

   a. Calculate the flight Mach number, M_∞ and the dynamic pressure, q_∞
   b. Calculate the flight Reynolds number based on its average wing chord
   c. Calculate the stagnation temperature in °R
   d. Calculate the pitot pressure in psfa using isentropic compressible flow equations
   e. At a point on the wing where V_1 = 600 ft/s, calculate T_1, p_1 and density, ρ_1.

4.8) A high-speed (use compressible flow equations) airplane moves at V_∞ = 250 m/sec through standard sea level air

   a. Calculate the Pitot tube total temperature, T_o, and total pressure, p_o.
   b. The airspeed gage is actually a pressure gauge which measures p_{gauge} = $(p_o - p_\infty)$ and is calibrated in m/sec by assuming sea-level density ρ_∞ = 1.225 kg/m³. This gauge reads what is called "calibrated" airspeed, V_{cal}. Calculate its value in this case.

4.9) A basketball with a 10 inch diameter drops through the hoop at a velocity of V = 35 ft/sec. Assuming sea-level air, calculate its Reynolds number Re (or N_R). Use Figure 4-34 of "experimentally evaluated drag coefficients for axis-symmetric bodies" to find the drag coefficient, C_D, of the ball, its dynamic pressure, q_∞ and drag, D.

4.10) Inside a boundary layer on an airplane, what is the ratio of the velocity at the plane's surface to that at the outer edge of the boundary layer at V_∞ = 295 ft/sec?

4.11) Calculate the thickness of the boundary layer at a location, x, on a flat plate in standard sea level air moving at 25 m/sec for the following two cases:

   a. x = 5 cm
   b. x = 2 m

4.12) A tennis ball serve is clocked at V_∞ = 175 ft/s (120 mph) in standard sea level air. Assuming the tennis ball diameter d = 0.2 ft...

   a. Calculate its free stream dynamic pressure q_∞ in psfa.
   b. Calculate the Reynolds number, based on its diameter. Use for viscosity: μ = 3.7373 x 10^{-7} slug/ft·sec.
   c. From a video analysis of its rate of deceleration, its drag is calculated to be D = 0.23 lb$_f$. Calculate its corresponding drag coefficient C_D.
   d. At this value of C_D, what type of boundary layer do you expect: laminar or turbulent?

CHAPTER 5. AIRFOILS AND WINGS

5.1 INTRODUCTION

Recall that most modern airplanes are made up of the following components:

1. The fuselage serves to connect components together like wings and tail control surfaces and house the pilot's cockpit, airplane controls and cargo or passengers.
2. The wing of reference area (S_w) provides lift (L) from a reduced air pressure over its upper surface. In steady level flight, lift is equal and opposite to weight (W).
3. An engine is needed to provide thrust (T), to compensate for airplane drag (D) in steady level flight. Extra thrust is needed for climb and acceleration during take-off.
4. The vertical and horizontal stabilizers are needed for pitch and yaw control. Together they make up the empennage which is installed at tail length (l_t), aft of the airplane's center of gravity. Its fixed surfaces provide the moments needed for stabile flight. Its movable elevator and rudder surfaces allow control in pitch and yaw.
5. Ailerons are needed for roll control to keep the wings level and to initiate and terminate a banked turn. They are installed near the wing tips at a large distance from airplane center of gravity.
6. A landing gear is needed for minimum drag take-off and landing and must be tall enough to provide propeller ground clearance. A retractable landing gear reduces drag in flight, thereby increasing cruise speed.

In 1799, George Cayley became the first person to recognize the need for most of the components found in a modern airplane. He realized that controlled flight requires separate components for lift, thrust, pitch and yaw control. His contemporaries were investigating flapping and rotary wings while Sir Cayley proposed a fixed wing for lift and rowing paddles for thrust. Cayley improved the wing geometry by testing and found that wing curvature increased wing performance over that of a flat surface. He suggested a vertical tail, like those of a fish for yaw control and a horizontal tail like that of a whale for pitch control. In 1804 Cayley flew a meter-long simple glider that performed and looked much likes a modern glider, shown below in Figure 5-1. In 1810, Cayley published a triple paper "On Aerial Navigation" in the Journal of Natural Philosophy.

Figure 5-1: The first modern airplane configuration depiction in history – Cayley's model glider, 1804.

Arguably, the most critical component of any aircraft is the collection of lift-generating surfaces which are incorporated into the overall aircraft design. The main wing(s) serve as the primary lifting surface(s), but canards may also be employed to generate lift. In either case, wings are simply three dimensional extrusions of special two dimensional forms known as airfoils.

5.2   TWO-DIMENSIONAL AIRFOILS

AIRFOIL BASICS AND BRIEF HISTORY

As early as the mid-nineteenth century it was known, though not well understood, that curved surfaces immersed in a flowing fluid would tend to generate a force considerably unaligned to the direction of the fluid flow. Depending upon the angle of the surface relative to the oncoming fluid, the component of force oriented normal to the flow (lift) was often an order of magnitude greater than the component of force aligned with the direction of the flow (drag) as shown schematically in Figure 5-2. The advantageously large lift-to-drag (L/D) relationship offered by curved surfaces, known collectively today as airfoils (also as wing sections or aerofoils) has driven more than a century of exhaustive studies of their various forms and has led to today's massive database of airfoils.

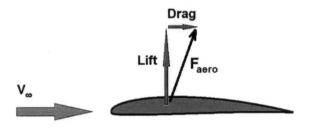

Figure 5-2: An airfoil immersed in a flowing fluid produces force unaligned with the direction of the flow.

There are as many reasons for specific airfoil geometry developments as there are purposes for airfoils themselves, but the methods used to obtain such variations have been relatively constrained. In their early years, airfoils were created based upon naturally occurring shapes such as the outline of brook trout swimming in the water below or the cross sections of bird wings as they piloted the skies above. Engineering pioneers like Leonardo da Vinci, Daniel Bernoulli, and Otto Lilienthal could see that in nature, special shapes allowed for the generation of this force. Lilienthal, for example, investigated a number of bird wings, dissecting them and recording their 2D shapes (Figure 5-3) for use in his "bird man" gliders.

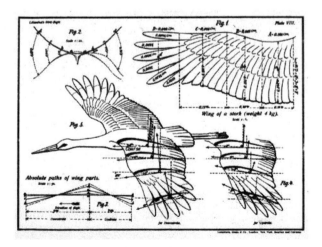

Figure 5-3: The flight mechanics and wing section shapes of a stork (Der Vogelflug als Grundlage der Fliegekunst, Otto Lilienthal, 1889).

These classic wing sections, in particular the Göttingen 398 and the Clark Y (Figure 5-4), ushered in the next phase of airfoil development in the United States by the National Advisory Committee for Aeronautics (NACA). NACA theoreticians, researchers and engineers systematically tested huge numbers of airfoils in two dimensional tests (i.e., tests relating to wings of infinite span and constant cross section) to ascertain the various functions of common critical geometric parameters and catalogue their performance over a wide range of relative flow angles, and operating Reynolds and Mach numbers (it is fair to point out that just as many three dimensional tests were also performed by the NACA, but these relate more to wing-specific geometric parameters). Simplistically speaking, accessing this great body of information once it had been developed meant only to have a specific aerodynamic need (e.g., a known operating Reynolds or Mach number range, cruise lift coefficient and related "drag bucket") which would take the wing designer to pages of airfoils, from which one especially fitting would be selected for the task.

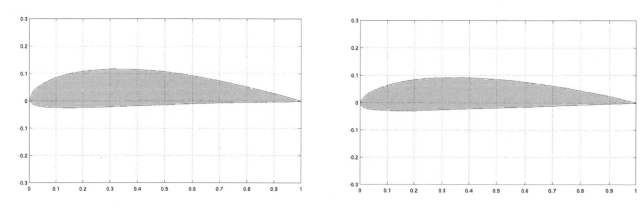

Figure 5-4: The Göttingen 398 (left) developed in 1919 and the Clark Y (right) developed in 1922 were used as the basis for a family of wing sections studied at the NACA in the 1920's.

It was during this illustrious period of NACA wing section development that the remarkable discoveries, evolving theories and analytical design approaches being made or refined were combined with practical computing power. Around the same time that the NACA had become the foundation for the National Aeronautics and Space Administration (NASA) in the late 1950's the last phase in the history of airfoil development began to emerge, enabled by the invention of the integrated circuit. The revolution of the microchip finally allowed aerospace engineers to practically approach the problem of airfoil design head-on by utilizing what is now known as "inverse design". This approach merely required the foreknowledge of a desired pressure coefficient distribution over a wing section to compute the external geometry required to produce such a pressure contour in a theoretical flow field. Today, this method is becoming increasingly easy to use, and is even available to the casual designer or hobbyist by way of freely downloadable software which can run on even a low-end processor. Additional computational methods still allow the "direct design" of airfoils – that is to say the trial and error method of designing the shape first and testing the performance afterward – and can promote the rapid evolution of wing section shapes without ever having to visit the wind tunnel or local airstrip for flight testing.

Airfoils need not be applied strictly to flight vehicles. Robert Liebeck, an aerospace engineer who has a long history of work with airfoils, has designed a series of airfoils for use specifically within the automotive racing industry. Liebeck airfoils are used on NASCAR and Formula One (F1) race cars. Because the airfoils are able to generate L/D ratios greater than 200, they can actually deliver as much as 3.5 times the weight of the car in aerodynamic downforce while the car is driving at top speeds. This force gives the cars much better cornering capability (from the added friction force between the tires and the racetrack), and, at least theoretically speaking, could allow these cars to drive at full speed upside-down. Examples of some of the F1 front wings are shown in Figure 5-5.

Figure 5-5: Front wings (sometimes called spoilers) on F1 cars are used to generate aerodynamic downforce to improve vehicle handling. *Left to right:* '78 Lotus, '79 Ferrari 312, '95 McLaren MP4-10.

ANATOMY OF THE AIRFOIL

Considering the form of a typical airfoil, there are several key components of the geometry which should be well-defined. The chord length, c, is the primary geometric value for the entire shape. The chord length is simply the distance between the forward-most part of the shape (the leading edge or LE) and the aft-most part of the shape (the trailing edge or TE). This distance can be represented by what is known as the chord line. It is important to note that the chord line is always the line of reference used in conjunction with freestream velocity direction to determine the airfoil angle of attack, α. The mean line of an airfoil is defined as the line for which any other line drawn normal to it and which extends from the lower surface to the upper surface will be bisected by the mean line. It is often extremely difficult to determine the mean line of an airfoil after it has been designed. Note also that for any thin airfoil, the mean line is the general predictor of the shape's aerodynamic behavior. Any airfoil for which the chord line and mean line are identical is known as a symmetric airfoil. Any airfoil which has distinctly different chord and mean lines is known as a cambered airfoil. A final critical feature of any airfoil is the section thickness, t. In general, for low speed airfoils, thicker is better, but at higher speeds, thinner is better. Note that all geometries, including thickness, are almost always provided in terms of chord length, thus a 10% thick airfoil is one for which the ratio of thickness to chord length is 1/10.

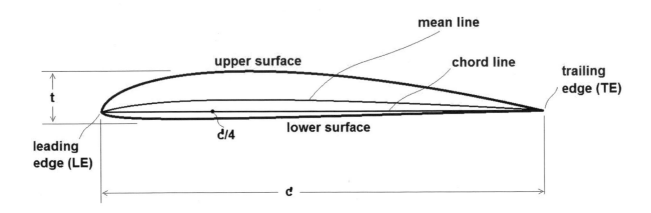

Figure 5-6: A cambered airfoil has a straight chord line and a curved mean line between the LE and TE.

DYNAMICS OF 2D AIRFOILS

It is prudent to consider how it is that the airfoil shape is capable of generating lift. It has already been shown that a back spinning cylinder placed in a flowing fluid will generate lift simply due to the pressure distribution around its surface as a result of the fluid being faster on the upper surface and slower on the lower surface. It happens that the two-dimensional lift generated due to the rotational motion of the cylinder is directly proportional to the circulation of the fluid around the cylinder. This circulation can be quantified as a vortex strength, Γ, which is calculated as the line integral of the velocity around a closed contour,

$$\Gamma = \oint \vec{V} \cdot d\vec{s} \qquad (5.1)$$

Careful evaluation of the units of vortex strength shows that it is a type of spanwise energy per unit circulating mass flow rate. More importantly, the lift per unit span, L', is directly related to Γ as

$$L' = \rho_\infty V_\infty \Gamma \qquad (5.2)$$

Known as the Kutta-Joukowski theorem, Eq. (5.2) is a famous relationship derived simultaneously by three mathematicians in the early 1900's (Frederick Lanchester in England, Wilhelm Kutta in Germany, and Nikolai Joukowski in Russia). The derivation of the theorem is beyond the scope of a typical introductory course in aerospace engineering, but it can be shown that the theorem applies to any closed two-dimensional body of arbitrary shape (Figure 5-7). In particular, one will note that the Kutta-Joukowski theorem is especially useful for predicting the 2D lift (i.e., lift per unit span) of airfoils.

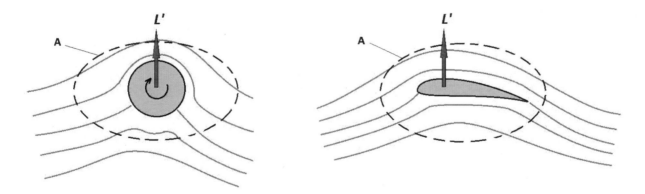

Figure 5-7: The strength of the circulation, Γ, about a closed 2D body of arbitrary shape (calculated by taking the line integral of the velocity field along the closed contour, A) may be used to calculate the lift per unit span, L', generated by back spinning cylinders (left) and airfoils (right).

The flow around an airfoil does not start with the fluid leaving the shape cleanly at the trailing edge. Fluid will always tend to depart from a surface at sharp edges (it may be helpful to envision water droplets leaving the ceiling of a damp cave from the pointy tips of stalactites, or to imagine hot air "dripping" up off the sharp peaks of small hills on the ground on hot, sunny days). This phenomenon as applied to airfoils is known as the Kutta condition. It is imperative for efficient use of airfoils that the trailing edge is as sharp as possible to encourage what is known as bound circulation on the surface of the shape, thus enhancing lift per unit span.

Examining the general behavior of a fluid around an airfoil in more detail, it is helpful to consider the velocity field around the shape, as shown schematically as streamlines in Figure 5-8.

Figure 5-8: Typical flow around an airfoil once the flow has fully "started". Note that the flow tends to exit the shape at the sharp trailing edge as enforced by the Kutta condition.

For airfoils (which are by definition two dimensional), the flow around the shape exhibits both upwash (i.e., flow in a relative upward direction) just at the LE and downwash (i.e., flow in a relative downward direction) just aft of the TE. This is not always the case for airfoil-based wings of finite span as will be shown later in this chapter. Using Bernoulli's equation, it is possible to sketch the type of static pressure field that must exist around the airfoil as a result of the flow depicted in Figure 5-8. This pressure distribution is shown schematically in Figure 5-9.

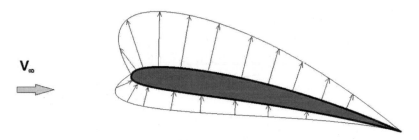

Figure 5-9: Typical static pressure distribution around an airfoil in a flow. Arrows indicate gage pressures (i.e., away from surface is negative gage pressure, inward into surface is positive gage pressure).

The overall result of the pressure integrated over the entire surface of the airfoil (as well as the shear stress exerted by the fluid on the surface) gives rise to three fundamentally important forces and moments on the airfoil: spanwise lift (L'), spanwise drag (D'), and spanwise pitching moment about the quarter-chord location ($M'_{c/4}$), shown in Figure 5-10.

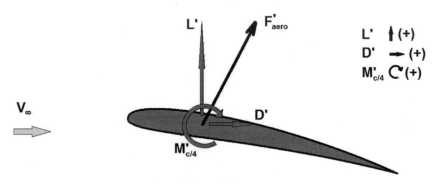

Figure 5-10: The variations in velocity, thus pressure distribution, over the airfoil give spanwise lift, drag and pitching moment about the quarter-chord (lift upward, drag in the direction of flow, and "nose up" pitching moment directions are all considered positive).

Airfoil forces and moment per unit span may be expressed as

$$L' = C_l q_\infty c \tag{5.3}$$

$$D' = C_d q_\infty c \tag{5.4}$$

$$M'_{c/4} = C_{M,c/4} q_\infty c^2 \tag{5.5}$$

It should be noted that while a nose-up pitching moment about the quarter chord of the airfoil is considered positive, most airfoils generate a negative pitching moment. This is often desirable for the airfoil to generate a restorative moment for angle of attack (i.e., for positive angles of attack, it is good to have a negative or nose-down pitching moment to restore the airfoil to a zero degree angle of attack). This behavior avoids dangerous tendencies for the airfoil to drive towards a stall condition for small positive angles of attack.

2D AIRFOIL TEST DATA

The incredible amount of experimental, numerical, and analytical work invested in understanding the spanwise force and moment behavior of 2D shapes – including that most intriguing and useful of shapes, the airfoil – is obvious in the volumes of data going back a hundred years or so. Having studied so many variations of the airfoils over this period of time, it is possible to make some important observations. In general, the lift and drag per unit span behavior of airfoils obey the following trends:

- o Typically, two dimensional lift (also called section lift) is linearly related to angle of attack.
- o Section drag is most often shown as a function of section lift and usually has a parabolic (quadratic) relation to two dimensional lift.
- o Section moment is usually nearly constant over the typical "useful" range of angle of attack.

One might wonder, "Why bother with 2D airfoil data? Why not just study 3D wings?" The answer is quite simple: the most useful way to be able to apply collected airfoil data to a specific condition is to adapt generic 2D data to specified 3D situations. Since there is an infinite number of wings which can be constructed from a single cross-sectional shape, it makes the most engineering sense to engage primarily in 2D studies assuming there is a smart way to translate the 2D information to a given 3D wing shape. This particular transformation process will be studied in detail later in this chapter. First, it would be wise to examine typical 2D airfoil data curves and try to understand the more important characteristics about them. Figure 5-11 presents two representative curves of primary importance for airfoils.

Examining the left half of Figure 5-11, 2D lift data is most usefully plotted as the two-dimensional lift coefficient, C_l, (note the lower case subscript "l") as a function of angle of attack, α. This type of plot is often referred to as the airfoil lift curve. For aircraft designers, the angle of attack is usually taken in terms of degrees, but for flight vehicle controls engineers, the angle of attack is typically taken in radians to make control law development less complicated. It is easy to see from the 2D lift curve that a typical airfoil generates lift as a linear function over a particular range of angle of attack. The range of angle of attack for which the lift curve remains linear with respect to angle of attack can be considered the "useful range" of the airfoil (i.e., the range within which the airfoil is not stalled). Over the useful range of the typical airfoil, the lift curve slope may be expressed as

$$C_{l_\alpha} = \frac{dC_l}{d\alpha} = a_0$$

The angle at which the section lift is zero is called the zero lift angle of attack, $\alpha_{L=0}$. Furthermore, at the angle where the lift suddenly falls with increasing angle of attack, the airfoil just begins to stall and the maximum section lift occurs at the stall angle, such that

$$C_{l,max} = C_l(\alpha_{stall})$$

It is possible to show the section lift over the useful range of angle of attack now as

$$C_l(\alpha) = C_{l\alpha}(\alpha - \alpha_{L=0})$$

Examining now the right half of Figure 5-11, 2D drag data is typically plotted as the section drag coefficient, C_d, (note the lower case subscript "d") as a function of C_l. This type of plot is often referred to as the airfoil drag polar. It is helpful in many cases to generalize the section drag coefficient as a parabolic function of section lift coefficient, such that

$$C_d = C_{d0} + kC_l^2$$

where C_{d0} is the minimum possible value of section drag coefficient and the constant, k, is found as

$$k = C_{d,C_l=1.0} - C_{d0}$$

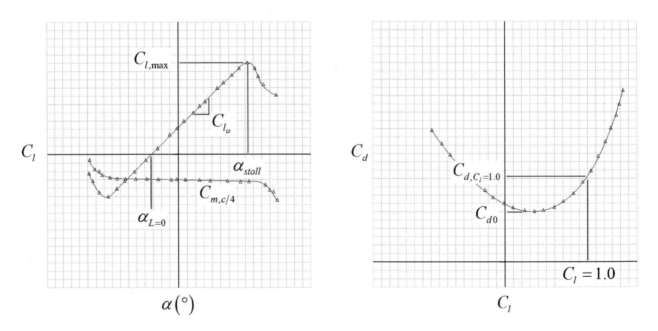

Figure 5-11: Typical section lift and quarter-chord pitching moment coefficient (left) and section drag coefficient (right) behavior.

NACA AIRFOILS

The National Advisory Committee for Aeronautics, or NACA, was formed in 1915 as an emergency response to the aviation necessities of World War I. In 1958, the NACA assets were essentially transferred over to what is known today as the National Air and Space Administration, NASA. The NACA's exhaustive study of a huge number of airfoil-based wings is partially culminated in *Theory of Wing Sections Including a Summary of Airfoil Data* by Ira Abbott and Albert von Doenhoff. In this tremendous work, one can find a literal treasure trove of airfoil information – everything from analytical methods of design to raw experimental section lift, drag, and pitching moment data. Many researchers have continued in the same spirit of the NACA as evidenced in such excellent resource websites as the one maintained by Michael Selig (of University of Illinois and Urbana Champaign or UIUC). A huge number airfoil coordinates and low speed data may be found at http://aerospace.illinois.edu/m-selig/ads.html.

As an example of the amazing amount of work that NACA engineers and scientists did, it is helpful to consider one of the simplest airfoil families the NACA studied – the famous NACA four-digit airfoils. In order to methodically check how the various airfoil shapes influenced the global behavior of an airfoil-based wing, researchers slowly modified the shapes according to a simple function. Each of the four digits represented some aspect of the airfoil shape:

- o The first digit represents the maximum difference between the mean line and chord line, a distance which is referred to as camber (in % c)
- o The second digit represents the x-location of the maximum camber (in 10% c)
- o The last two digits represent the thickness of the airfoil (in % c)

So, the NACA 2412 airfoil is a 12% thick airfoil which has a 2% camber located 40% aft of the LE.

One might reasonably be wondering how it is even possible to perform 2D tests. In reality, there is no such thing as a 2D wind tunnel test model, but it is possible to "trick" the flow moving over the wing into thinking the wing has infinite span (and is thus 2D in nature) by extending the wing all the way across the wind tunnel test section from wall to wall. If it is not possible to do that, then endplates can also be installed on the wing tips to effectively create the same kind of flow condition. Both approaches are presented graphically in Figure 5-12.

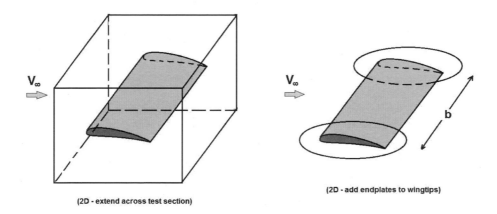

(2D - extend across test section)

(2D - add endplates to wingtips)

Figure 5-12: Methods of performing 2D tests in a wind tunnel include spanning the entire test section or installing wingtip endplates.

With either of these two arrangements, the real measured force is collected with the understanding that the result is really 2D in nature, thus the forces or moment divided by the span, b, gives the 2D data:

$$L_{measured} = C_l q_\infty cb \qquad \Rightarrow \qquad L' = \frac{L_{measured}}{b} = C_l q_\infty c$$

$$D_{measured} = C_d q_\infty cb \qquad \Rightarrow \qquad D' = \frac{D_{measured}}{b} = C_d q_\infty c$$

$$M_{c/4,measured} = C_{m,c/4} q_\infty c^2 b \qquad \Rightarrow \qquad M'_{c/4} = \frac{M_{c/4,measured}}{b} = C_{m,c/4} q_\infty c^2$$

MEASURING 2D LIFT AND DRAG USING PRESSURE

The most common method of measuring lift and drag from wing sections is to use strain gage-based load cells that can provide voltage differential output that can be correlated to applied force loads. One of the complications that arises from this method is that it becomes extremely difficult, if not impossible, to remove the effects of drag from artificial components such as endplates from the true drag of the airfoil itself. Additional complications of drag stem from the presence of walls or endplates nearby the airfoil-based wing (recall a portion of parasitic drag comes from interference effects).

Fortunately there is another more fundamental way of actually measuring 2D forces and moments. This is done using wing surface static pressure taps aligned on a single cross-sectional plane of a 2D-oriented wind tunnel model as well as using what is known as a wake rake downstream of the model behind the section region with the pressure taps. A series of small tubes connect to the airfoil static pressure taps and ports on the wake rake to pressure measuring devices such as a simple bank of U-tube manometers.

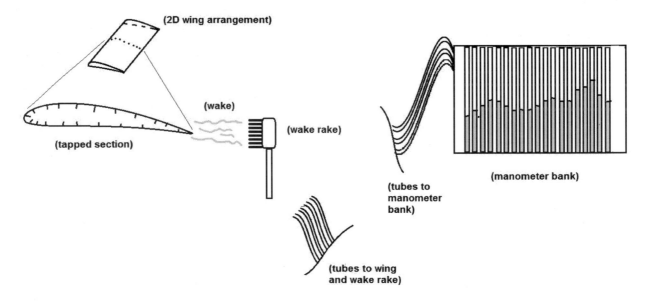

Figure 5-13: Static taps on a single plane of an airfoil-based wing and pressure ports on a downstream wake measuring probe called a wake rake are often used to measure section lift and drag.

The pressure taps provide all the insight into the pressure related lift and drag generated by the wing section, and the wake rake also allows researchers the opportunity to evaluate any momentum deficit within the wake region of the wing section. These two sources combined allow for a detailed 2D understanding of the wing section 2D aerodynamics. To allow the results to be as broadly applicable as possible, the pressure terms are often made dimensionless as a pressure coefficient, C_p, such that

$$C_p = \frac{p - p_\infty}{q_\infty} \qquad (5.6)$$

Note that for any location of relative suction (high local velocity), the value of the pressure coefficient is negative, and for any location of relative high pressure (low local velocity), the pressure coefficient is positive. The maximum possible pressure coefficient on an airfoil occurs wherever the flow has stagnated (i.e., velocity is zero), and is readily found as

$$C_{p,\max} = 1$$

The minimum value of the pressure coefficient is similarly found as

$$C_{p,\min} = 1 - \frac{p_0}{q_\infty}$$

EXAMPLE 5.1: 2D LIFT, DRAG, AND PITCHING MOMENT FOR A NACA 4415-BASED WING

Given:

A wing based on the NACA 4415 is subjected to 2D tests in a wind tunnel under the following conditions:

$$NACA\,4415$$

$$V_{test} = 680\,ft/\sec \qquad \rho_{test} = 0.00220\,slug/ft^3 \qquad \mu_{test} = 3.7373 \times 10^{-7}\,slug/ft - \sec$$
$$b = 2\,ft \qquad\qquad c = 9\,in \qquad\qquad \alpha_{test} = 6°$$

Find:

a) The operating Reynolds number of the wing
b) The lift generated by the wing
c) The drag generated by the wing
d) The pitching moment generated by the wing
e) The stall angle and the lift-to-drag ratio of the wing

Solution:

a)

$$\text{Re}_c = \frac{\rho Vc}{\mu} = \frac{0.0022\,slug}{ft^3} \left| \frac{680\,ft}{\sec} \right| \frac{9\,in}{12\,in} \left| \frac{ft}{3.7373 \times 10^{-7}} \right| \frac{ft - \sec}{} = 3 \times 10^6$$

b)

$$L' = C_l q_{test} c$$

from graph, at $\alpha = 6°$, $C_l \approx 1.0$

$$q_{test} = \tfrac{1}{2} \rho_{test} V_{test}^2 = \frac{1}{2} \left| \frac{0.0022\,slug}{ft^3} \right| \frac{680^2\,ft^2}{\sec^2} \left| \frac{lb_f - \sec^2}{slug - ft} \right| = 509\,lb_f/ft^2$$

$$L' = \frac{1.0}{} \left| \frac{509\,lb_f}{ft^2} \right| \frac{9\,in}{} \left| \frac{ft}{12\,in} \right| = 381\,lb_f/ft$$

c)

$$D' = C_d q_{test} c$$

from graph, at $C_l = 1.0, C_d \approx 0.008$

$$D' = \frac{0.008}{} \left| \frac{509\,lb_f}{ft^2} \right| \frac{9\,in}{} \left| \frac{ft}{12\,in} \right| = 3.05\,lb_f/ft$$

d)

$$M'_{c/4} = C_{m,c/4} q_{test} c^2$$

from graph, at $\alpha = 6°$, $C_{m,c/4} \approx -0.08$

$$M'_{c/4} = \frac{-0.08}{ft^2} \left| \frac{509\,lb_f}{ft^2} \right| \frac{9^2\,in^2}{} \left| \frac{ft^2}{12^2\,in^2} \right. = -22.9\,ft - lb_f / ft$$

e)

$$\alpha_{stall} \approx 14° \qquad \left(Note: C_{l,max} \approx 1.62 \right)$$

$$L/D = \frac{L'}{D'} = \frac{381}{3.05} = 125! \qquad \leftarrow \qquad \text{That's quite good!}$$

Check:

The values seem to be of appropriate magnitude and sign, and have the correct units.

5.3   THREE-DIMENSIONAL WINGS

WING FLOW BASICS

The question of how to implement 2D airfoil shapes in a 3D application is of great concern to the aerospace design engineer. Fortunately, there are well known and accepted methods which allow engineers to translate 2D data from experimental, computational, or analytical methods to appropriate airfoil applications. Finite wings make up the vast majority of airfoil applications. It is absolutely critical to understand a very fundamental difference between airfoils and airfoil-based wings: 3D airfoil based wings have a finite span, *b*, whereas airfoils are essentially infinite in span. It is therefore reasonable to draw the valid conclusion that *any aerodynamic difference between 2D airfoils and 3D airfoil-based wings has everything to do with the existence (or lack thereof) of wingtips.* Consider a finite wing in an oncoming flow as shown in Figure 5-14:

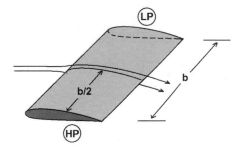

Figure 5-14: A finite airfoil-based wing of span, *b*, immersed in an oncoming fluid flow.

At the semispan the relative high pressure (HP) acting on the lower surface and the relative low pressure (LP) acting on the upper surface can only "see" one another and interact at the wing LE and TE. In some sense, the flow along the semispan has some notion of the presence of the wingtips, but due to symmetry, it moves in a straight fashion along the upper and lower surfaces and does not exhibit any spanwise component of its velocity. At any other spanwise location on the wing, however, there is an obvious preferred direction of flow due to local pressure gradients which the fluid will tend to follow. Only static pressure gradients drive flows, so consider what happens to the fluid flow as a result of the local pressure gradients. Figure 5-15 presents four views to help understand how fluid flows over an airfoil-based wing at a positive angle of attack.

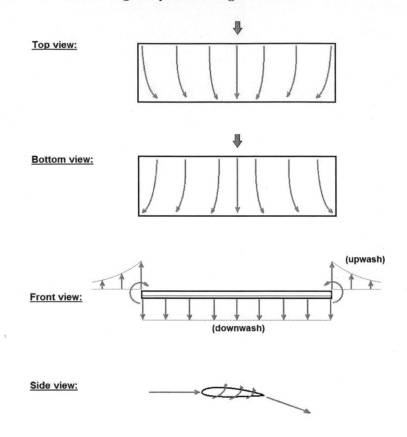

Figure 5-15: Pressure gradients around the airfoil-based wing drive the movement of fluid around it.

LIFT INDUCED DRAG AND ASPECT RATIO

In life there is usually no such thing as a free lunch. Real world lift-generating surfaces are no exception. One extremely unfortunate side effect of finite wing lift generation is the phenomenon known as "lift induced drag", or more commonly as just "induced drag". It turns out that whenever a wing of finite space is used to generate lift, it also necessarily generates an appreciable amount of force opposing the forward motion of the wing. Careful inspection of Figure 5-15 helps to show why this is the case: as relatively high pressure air is "bled" around the wing tips into the lower pressure region, moving from the lower surface to the upper surface, a wing tip vortex is formed at each end of the wing. The presence of the pair of wingtip vortices influences the wake region of the wing in such a way as to cause a general downward motion of the entire region of flow directly aft of the wing. This downward motion, coupled with the general freestream velocity, causes an effective velocity which is rotated slightly downward. This is shown pictorially in Figure 5-16.

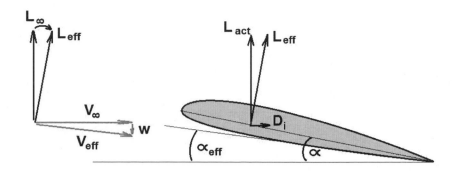

Figure 5-16: A finite wing that generates lift causes a downwash velocity that changes the effective velocity experienced by the wing.

Lift is, by definition, that component of aerodynamic force which operates normal to the effective velocity surrounding an object. The fact that the effective velocity for a finite wing generating lift has a slight downward angle causes the lift vector to rotate aft. Since the actual direction of travel is still aligned with V_∞, a component of lift necessarily opposes the direction of motion. This aft-pointing component of lift is called lift-induced drag, or induced drag, D_i, for short. Fortunately, the induced drag is fairly easy to calculate as

$$D_i = C_{Di} q_\infty S_w \tag{5.7}$$

where

$$S_w = bc \tag{5.8}$$

and

$$C_{Di} = \frac{C_L^2}{\pi e A R} \tag{5.9}$$

It is relatively easy to show how Eq. (5.9) may take its form. Using Figure 5-17 as a reference, consider the mass flow rate of a fluid such as air entering normal to the disk region defined as a region bisected by a lift-generating wing; the face of the disk region is oriented normal to the freestream flow of air. As a result of the presence of the wingtip vortices, the incoming mass flow rate, along with its associated momentum, is directed slightly downward.

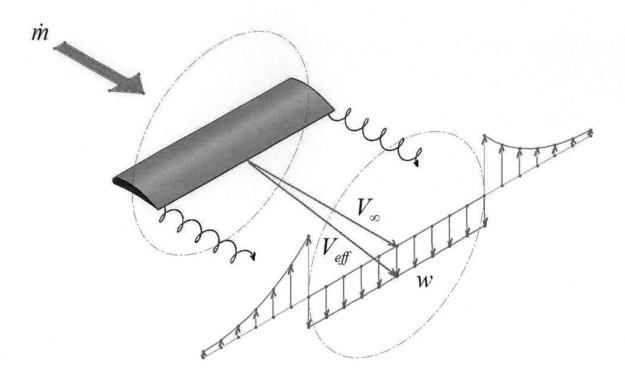

Figure 5-17: Wingtip vortices induce a downward flow aft of the wing resulting in an effective velocity which is rotated slightly downward.

According to Newton's second law, this change in momentum must be equivalent to the lift generated by the wing, or

$$\frac{d}{dt}(mV) = \dot{m}w = L$$

Noting that the mass flow rate entering the disk for a wing of span, b, is found as

$$\dot{m} = \rho A V = \rho_\infty V_\infty \frac{\pi b^2}{4}$$

and solving for downwash,

$$w = \frac{4L}{\rho_\infty V_\infty \pi b^2}$$

The power required to accelerate this mass downward may be called the induced power, P_i, where

$$P_i = D_i V_\infty$$

But the induced power is also the time rate of change of kinetic energy within the flow, or

$$\frac{d}{dt}(KE) = P_i = \frac{d}{dt}\left(\frac{mw^2}{2}\right) = \frac{\dot{m}w^2}{2} = \frac{Lw}{2}$$

Thus

$$D_i = \frac{Lw}{2V_\infty} = \frac{L^2}{\frac{1}{2}\rho_\infty V_\infty^2 \pi b^2} = \frac{L^2}{q_\infty \pi b^2}$$

Noting that

$$L^2 = \left(C_L q_\infty S_w\right)^2$$

and defining the induced drag coefficient as

$$C_{Di} = \frac{D_i}{q_\infty S_w}$$

one may write

$$C_{Di} = \frac{C_L^2 S_w}{\pi b^2}$$

As pointed out in earlier chapters, aerodynamicists often choose to work with dimensionless variables which assist in understanding certain critical relationships. Consider one such dimensionless variable, known as aspect ratio, AR, which compares the wing area and span such that

$$AR = \frac{b^2}{S_w} \tag{5.10}$$

Using Eq. (5.10), it is finally possible to write

$$C_{Di} = \frac{C_L^2}{\pi AR}$$

Which is almost identical to Eq. (5.9), with the exception of the variable, e, in the denominator. The value e, known as "Oswald's efficiency factor", named after aerodynamicist W. B. Oswald of the Douglas Aircraft Company, who first used the term "efficiency factor", provides a way of evaluating the spanwise wing loading efficiency of the wing. This will be expanded upon in more detail in the coming sections.

Before moving on, two important notes must be made. First, it is important to note that in the limit,

$$\lim_{AR \to \infty} \left(C_{Di}\right) = 0$$

In other words, as the span of a wing increases with respect to its area, the value of the induced drag coefficient decreases. Ultimately, a two dimensional wing, which is essentially a wing of infinite span, thus has no lift induced drag associated with it.

The second important thing to note is that if one examines Eq. (5.10) carefully, it will become obvious that for a rectangular wing (i.e., one for which the chord length is constant along its span and thus has a rectangular planform), aspect ratio is simply

$$AR_{rect} = \frac{b}{c} \tag{5.11}$$

For high aspect ratio wing, its general appearance will be slender in nature and its induced drag will tend to be relatively small. Conversely, a low aspect ratio wing has a generally stubby appearance and its induced drag will tend to be relatively high.

SPANWISE LOADING, WING INCIDENCE, TWIST, AND TAPER

As previously noted, e is known as the Oswald spanwise load efficiency factor. This term provides a way of estimating the ability for a wing to generate as much lift as it may be theoretically able to do. In basic terms, there is a significant amount of potential lift which is lost due to pressure "spillage" over the wingtips, and this loss can be mitigated if the spanwise lift distribution is managed in some clever way. It turns out that the most efficient way to load a wing is to have an elliptic spanwise load distribution, which may be described as having a 100% loading efficiency, or an Oswald's efficiency of unity. Empirically, all other loading efficiencies may be estimated as

$$e = \frac{2}{\left[2 - AR + \sqrt{\left(4 + AR^2 \right)} \right]}$$

Note that in the limit,

$$\lim_{AR \to \infty} \left(e \right) = 1$$

Consider two planforms, one elliptic and one rectangular, both with identical wing area. The lift loss at the tips is worse for the rectangular planform, resulting in an overall loss of wing lift as compared to the elliptic wing.

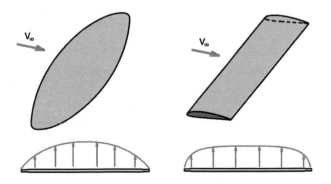

Figure 5-18: Elliptical (left) and rectangular (right) planforms exhibit different lift distributions.

The basic idea behind elliptically loading the wing is to gently reduce the pressure differential between the upper and lower surfaces of the wing near each of the wingtips. This has the effect of reducing the overall lifting potential loss. There are three basic ways to get a finite wing to have something like an elliptic load distribution:

1) Make the planform elliptic
2) Gradually taper the wing so that the root chord length is larger than the tip chord length
3) Moving from root to tip, gradually twist the airfoil shape in a nose-down direction

Elliptic planforms such as the one used on the British Spitfire are effective, though not very common because of the difficulty of wing fabrication. It is interesting to note, however, that even the Spitfire's wing was not designed to maximize lift distribution. Rather, the Spitfire's planform was designed to enable the use of three Browning .303 machine guns on board each wing. A tapered wing was preferred by the original designers, but there was not enough length on the outboard portions of the wings to get all three guns on each side without changing the planform to an elliptic shape. Taper and twist are used much more frequently to produce favorable spanwise load distributions. In particular, the taper ratio, λ, is given as

$$\lambda = \frac{c_t}{c_r}$$

where c_t is the chord length of the wing at its tip and c_r is the chord length of the wing at its root. Furthermore, it can be shown that the optimum taper ratio for maximum lift distribution is given as

$$\lambda_{opt} = 0.4$$

To establish the operating Reynolds number for a wing, it would certainly be easier to establish some average chord length for the wing by what is known as the mean aerodynamic chord, MAC, which for a tapered wing is found as

$$MAC = \left(\frac{2}{3}\right) c_r \left(\frac{1 + \lambda + \lambda^2}{1 + \lambda}\right)$$

2D AIRFOIL TO 3D WING PERFORMANCE CONVERSIONS

Aircraft design engineers need to be able to make reliable performance estimations for the wings that they design before the designs go to production. From the vast amount of 2D airfoil data available, it is possible to modify the known 2D values using known 3D wing specifications. Airfoil-based wing lift and drag performance are of special importance. To derive a governing equation for wing lift coefficient, first consider the 3D lift curve slope, a, over its linear range such that

$$a = \frac{dC_L}{d\alpha} = \frac{a_0}{\left(1 + \dfrac{57.3 a_0}{\pi e AR}\right)}$$

Because the wing lift is linear over the "useful range" of angle of attack, it is possible to express the wing lift coefficient, C_L, in terms of angle of attack as

$$C_L = a\left(\alpha - \alpha_{L=0}\right)$$

Where for a given airfoil-based wing

$$\alpha_{L=0} = \alpha_{L=0,3D} = \alpha_{L=0,2D} = const.$$

Now consider the wing drag coefficient. Aerodynamic drag on a finite wing is comprised of both parasitic drag and lift-induced drag, thus the total drag coefficient on a wing may be expressed as

$$C_D = C_{D,para} + C_{Di}$$

where

$$C_{D,para} = C_{D0,wing} = C_{d0}$$

Thus

$$C_{D,wing} = C_{d0} + C_{Di} = C_{d0} + \frac{C_L^2}{\pi e AR}$$

For simple rectangular extruded wings (i.e., a wing with neither twist nor taper), several additional aspects related to the conversion of 2D to 3D airfoil-to-wing data are based on the following points:

- Wing pitching moment is not terribly dependent on the existence of wingtips, so

$$C_{m,c/4,3D} \approx C_{m,c/4,2D}$$

- The angle of zero lift is completely independent of three dimensionality, so

$$\alpha_{L=0,3D} = \alpha_{L=0,2D}$$

- The maximum lift coefficient that a 3D wing is capable of generating varies only slightly from that of its 2D airfoil counterpart, although the angle of attack at which the maximum lift occurs does increase in the case of the wing, so in general

$$C_{L,\max} \approx C_{l,\max}$$
$$\alpha_{stall,3D} > \alpha_{stall,2D}$$

TEST YOUR UNDERSTANDING: ESTIMATING 3D LIFT COEFFICIENT FROM 2D DATA

A symmetric airfoil is tested in a wind tunnel and is found to have a lift coefficient of 0.93 at $\alpha = 9°$.

If a rectangular wing with a span of 7 feet and a chord length of 9 inches is to be made based on the symmetric airfoil and the anticipated Oswald's efficiency will be 0.88, what lift coefficient should the 3D wing be capable of generating at $\alpha = 9°$?

Answer: $C_L(9°) \approx 0.75$

EXAMPLE 5.2: FINDING 3D PERFORMANCE BASED ON 2D DATA

Consider the case of a tapered wing based on a Clark-Y airfoil:

Given:

Clark-Y Airfoil 2D Data for $Re_c = 6.0 \times 10^4$:

		3D Operating Parameters for Rectangular Wing:	
$\alpha_{L=0} \approx -1°$	$C_{d,0} \approx 0.027$	$c = 0.05\,m$	$q_\infty = 188\,Pa$
$C_{l,max} \approx 1.22$		$b = 0.3\,m$	$e = 0.85$
$C_l(9°) \approx 0.95$		$\alpha = 4°$	

Find:

a) The values of the 2D and 3D lift curve slopes
b) The value of the 3D wing lift and drag

Solution:

a)

$$a_0 = \frac{dC_l}{d\alpha} = \frac{\Delta C_l}{\Delta \alpha} = \frac{(0.95-0)}{(9-(-1))°} = 0.095\,(1/°)$$

$$a = \frac{dC_L}{d\alpha} = \frac{a_0}{\left(1+\dfrac{57.3a_0}{\pi e AR}\right)}; \qquad AR_{rect} = \frac{b}{c} = \frac{0.3\,m}{0.05\,m} = 6$$

$$a = \frac{0.095\,(1/°)}{\left(1+\dfrac{57.3(0.095)}{\pi(0.85)(6)}\right)} = 0.0709\,(1/°)$$

b)

$$L = C_L q_\infty S_w; \qquad C_L = a(\alpha - \alpha_{L=0}) = 0.0709\,(1/°)(4+1)° = 0.355$$

$$S_w = bc = (0.3\,m)(0.05\,m) = 0.015\,m^2$$

$$L = (0.355)(188\,Pa)(0.015\,m^2) = 1.00\,N$$

$$D = C_D q_\infty S_w = (C_{D,para} + C_{Di})q_\infty S_w = (C_{d,0} + C_{Di})q_\infty S_w$$

$$C_{Di} = \frac{C_L^2}{\pi e AR} = \frac{0.355^2}{\pi(0.85)(6)} = 0.00786$$

$$D = (0.027 + 0.00786)(188\,Pa)(0.015\,m^2) = 0.098\,N$$

Check:

The calculated values have the correct units and seem to have appropriate magnitude and sign.

WING CRUISE VELOCITY

As usual, engineers always try to find an optimized operational condition for which the minimum amount of waste or maximum efficiency is achieved in a selected system. The careful engineer will note, however, that this definition says nothing about how to select the system itself, thus the "optimum" condition for the wing designer may not be exactly the same as the optimum condition from the propulsion system designer's perspective. In the case of aircraft, the condition for which the flight vehicle operates at a minimum overall drag is the most commonly preferred operating condition. The speed at which this occurs is referred to as the cruise velocity.

In many cases, engineers will try to develop simplistic functional relationships between critical operational parameters. Applied to the concept of drag from a lifting-generating body, it behooves the diligent aerospace engineer to use some simple functional analysis of the various drag sources to determine the general operating restrictions associated with winged flight. Begin by examining the relationship of each drag component to flight velocity. For subsonic flows, recall from Eq. (4.27),

$$D = D_{para} + D_i$$

Investigating each component of drag with special emphasis on flight speed,

$$D_{para} = C_{D,para} q_\infty S_w = C_{d,0}\left(\tfrac{1}{2}\rho V_\infty^2\right) S_w \qquad \Rightarrow \qquad D_{para} \propto V_\infty^2$$

and

$$D_i = C_{Di} q_\infty S_w = \frac{C_L^2}{\pi e AR}\left(\tfrac{1}{2}\rho V_\infty^2\right) S_w = \frac{L^2}{\pi e AR\left(\tfrac{1}{2}\rho V_\infty^2\right) S_W} \qquad \Rightarrow \qquad D_i \propto V_\infty^{-2}$$

These relationships reveal an interesting functional relationship of overall aircraft drag with respect to operating speed, as shown in Figure 5-19.

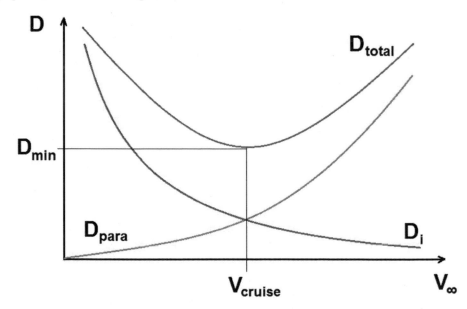

Figure 5-19: Minimum drag occurs at cruise velocity when parasitic and lift-induced drag are equivalent.

Note that as the flight velocity increases, the parasitic drag increases, but the induced drag decreases. This situation creates a distinct minimum drag value at the flight velocity where parasitic and induced drag are equivalent values, so that

$$D = D_{min} \quad \Rightarrow \quad D_{para} = D_i \quad \Rightarrow \quad C_{D,para} = C_{Di}$$

The velocity at which this minimum drag occurs is referred to as the cruise velocity, and represents the flight speed at which the aircraft performance has been maximized. Considering the aircraft lift-to-drag ratio as *the* performance parameter, and noting that for steady level flight the aircraft lift is equal to its weight, then

$$\left(\frac{L}{D}\right)_{max} = \frac{W}{D_{min}} \tag{5.12}$$

From this simple relationship, a host of optimized conditions are now easily calculated. For example, because the parasitic drag and the lift-induced drag are equivalent at cruise speed, their respective coefficients must also be equivalent, or

$$C_{Di,cruise} = \frac{C_{L,cruise}^2}{\pi eAR} = C_{D,para,cruise}$$

Thus, the optimal cruise lift coefficient may be related directly to the parasitic drag as

$$C_{L,opt} = \sqrt{\left(C_{D,para}\pi eAR\right)} \tag{5.13}$$

Note also that for steady, level flight at cruise speed,

$$C_{L,opt} = \frac{W}{\frac{1}{2}\rho V_{cruise}^2 S_w} \tag{5.14}$$

It is fascinating to note that for nearly all flying things, biological or manmade, large or small, there exists a striking similarity across the board in terms of optimum lift coefficient. It would seem from any list of common flyers that the typical lift coefficient at which the animal of vehicle most operates with the least drag tends to be at a value near 0.65. This fact will often help aerospace engineers, regardless of their level of experience, quickly evaluate potential aircraft design parameters: if the desired cruise speed, weight, wing area and operating density do not yield a value in the neighborhood of 0.65 when applied in Eq. (5.14), there may be problems in the design.

TEST YOUR UNDERSTANDING: FINDING OPTIMUM LIFT COEFFICIENT FOR A KESTREL

A kestrel (*falco tinninculus*) flies at the following conditions in standard sea level air:

$$V = 8.7\,m/sec \qquad W = 2.03\,N \qquad S_w = 0.0708\,m^2$$

Assuming the bird flies at the most efficient speed, what is its optimum lift coefficient?

Answer: $C_{L,opt} = 0.618$

WING STALL VELOCITY

Aside from cruise speed, there is one other critical operating speed which must be considered for any aircraft. It has already been shown that there is an angle of attack above which an airfoil or an airfoil-based wing will simply not be able to maintain attached flow on the upper surface of the structure. This condition is known as stall. In later chapters, the importance of stall velocity will be further investigated in designing aircraft for reasonable takeoff and landing distances. For now, recall that just before stall, the airfoil or airfoil-based wing exhibits its maximum section lift or lift coefficient, respectively. It follows that for a fixed lift value, there is an associated stall dynamic pressure, such that

$$q_{stall} = \frac{W}{C_{L,max} S_w}$$

(5.15)

Also note that the stall dynamic pressure may be further reduced to indicate the minimum flight speed required to maintain lift. This value is known as the stall velocity and is found as

$$V_{stall} = \sqrt{\frac{2q_{stall}}{\rho}}$$

(5.16)

Finally, Eq. (5.15) and Eq. (5.16) may be combined to find the stall velocity as

$$V_{stall} = \sqrt{\frac{2W}{C_{L,max} \rho S_w}}$$

(5.17)

Examining Eq. (5.17) in more detail reveals the main parameters involved in ensuring that aircraft can generate the kind of lift necessary to overcome weight and take to the air:

- The higher the aircraft weight, the higher the stall velocity. All other things being equal, a heavy Boeing 747 will have to go much faster than a light Piper Cub to get into the air.
- The lower the maximum lift coefficient the aircraft is capable of generating, the higher the stall velocity will be. All other things being equal, an aircraft that can generate higher maximum lift coefficients will get off the ground at much lower speeds.
- The lower the ambient density, the higher the stall velocity. All other things being equal, an aircraft taking off from a high mountain top airport will have a much higher stall speed than the same aircraft at sea level.
- The lower the wing area, the higher the stall velocity. Large wings mean low stall velocities.

TEST YOUR UNDERSTANDING: MAXIMUM LIFT COEFFICIENT FOR A BOEING 747-400

A Boeing 747-400 takes off with a maximum take-off weight of 875,000 lb$_f$. The aircraft has a wing area of 5,650 ft^2 is able to get airborne by the time it reaches the end of a two-mile runway flying at just 200 ft/sec.

Assuming operation at a sea level airport on a standard day, what is the maximum lift coefficient that the aircraft can generate to achieve take-off?

Answer: $C_{L,max}$ = 3.26

EXAMPLE 5.3: CALCULATING STALL AND CRUISE SPEEDS FOR A KNOWN WING

Consider the case of a simple wing with the following information:

Given:

$$NACA\,23012-based\,wing$$

$a = 0.078\,(1/°)$	$AR = 8.2$	$W = 7{,}100\,N$
$C_{L,max} = 1.6$	$S_w = 14\,m^2$	$\rho_\infty = \rho_{sl,std}$
$C_{D0} = 0.0065$	$e = 0.73$	

Find:

a) The wing cruise velocity
b) The wing stall velocity

Solution:

a)

$$\text{"cruise"} \Rightarrow C_D = C_{D,min} \Rightarrow C_{Di} = C_{D,para} = C_{D0}$$

$$C_{L,opt} = \sqrt{C_{D0}\pi e AR} = \left(0.0065 \times \pi \times 0.73 \times 8.2\right)^{\frac{1}{2}} = 0.349$$

$$q_{opt} = \frac{W}{C_{L,opt}S_w} = \frac{7100\,N}{\left(0.349 \times 14\,m^2\right)} = 1{,}451\,Pa$$

$$V_{cruise} = V_{opt} = \sqrt{\frac{2q_{opt}}{\rho_\infty}} = \left(\frac{2\times 1451\,Pa}{1.225\,kg/m^3}\right)^{\frac{1}{2}} = \boxed{48.7\,m/\sec}$$

b)

$$\text{"stall"} \Rightarrow C_L = C_{L,max} = 1.6$$

$$q_{stall} = \frac{W}{C_{L,max}S_w} = \frac{7100\,N}{\left(1.6 \times 14\,m^2\right)} = 317\,Pa$$

$$V_{stall} = \sqrt{\frac{2q_{stall}}{\rho_\infty}} = \left(\frac{2\times 317\,Pa}{1.225\,kg/m^3}\right)^{\frac{1}{2}} = \boxed{22.7\,m/\sec}$$

Check:

The calculated values have the correct units and seem to have appropriate magnitudes.

5.4   HIGH-LIFT DEVICES

A simple calculation of the maximum lift coefficient required to enable a typical large aircraft to take to the air might alarm the beginning aerospace engineer. How is it possible that airplane wings made from airfoils with maximum lift coefficients around 1.5 or less can be utilized in these situations? The short answer is they cannot be used without some form of modification. Passive geometric modifications (e.g., slats and flaps) as well as active lift-enhancing systems (e.g., blowing and suction devices) can be implemented into aircraft wings to affect such high lift performance.

For take-off on paved runways in minimal distance, one usually applies about 20 degrees of flap deflection to reduce tire weight and rolling friction, thus reducing the take-off speed, which is usually taken to be 120% of the stall velocity. Soft runways like grass require even more flap deflection. There are many kinds of flaps in use to increase C_{Lmax} and thereby reduce the stall speed. Very high flap deflections are only used when the associated high drag is acceptable. This is during landing on a limited runway length. Special STOL (Short Take-off and Landing) aircraft are designed for short runways. They incorporate additional high lift devices such as a drooped leading edge (LE), LE Krueger flap or LE slats. Some of these high lift devices are shown in Figure 5-20.

Fowler and other multiple element flaps delay stall till higher C_L values. This can be understood when considering what makes an airfoil stall. Prior to stall the air velocity over the upper surface reaches a peak near the nose of the airfoil. At that location the pressure will be at its lowest value. For a single-element airfoil the pressure at the trailing edge (TE) equals the ambient pressure, thus the upper surface boundary layer must have enough momentum to sustain an adverse pressure rise from the minimum value all the way back to ambient pressure at the TE. This limits the value of C_{Lmax}, especially at low Reynolds numbers, where the low turbulence level cannot prevent TE stall. By adding a separate element (e.g., Fowler flap) located just aft and below the trailing edge of the upstream one, the suction peak on the nose of the second element extends into the TE of the upstream element. This reduces its adverse pressure gradient, and allows it to operate at increased C_{Lmax}. Further, the airfoil lift coefficient is always calculated in reference to the cruise chord length. The extension of the effective chord and thus wing area, using a multiple element flap, further contributes to the increased value of C_{Lmax}. Chicago O'Hare has one runway of 10,000 ft length, which is relatively short for commercial aircraft, which can be very heavy. Note a fully loaded Boeing 747 weighs 750,000 pounds. Therefore commercial aircraft are equipped with high lift devices to maximize C_L and then minimize the required runway length.

Leading edge slats were used extensively in World War I aircraft. They prevent LE stall, which is much more dangerous than TE stall. This is because TE stall advances slowly upstream and thus gradually reduces the affected wing area. However LE stall instantaneously separates the flow over most of the upper surface with a significant reduction in lift. Especially in the dogfights of slow moving aircraft, LE stall is a common problem. Modern fighters handle this problem by using what is known as a "strake". The strake is a highly swept plate along the fuselage which produces a strong LE vortex and a high vacuum above it to hold the nose of the aircraft up. The LE slat works totally differently than the Fowler flap. It does not increase the lift on the wing. But because of its airfoil shape it does produce lift. Its circulation reduces the velocity in the gap between it and the LE of the main wing. This in turn reduces main wing peak velocity and vacuum on the nose of the main wing, thus preventing its L.E. from stalling. Note the lift produced by the slat equals the lift lost by the main wing.

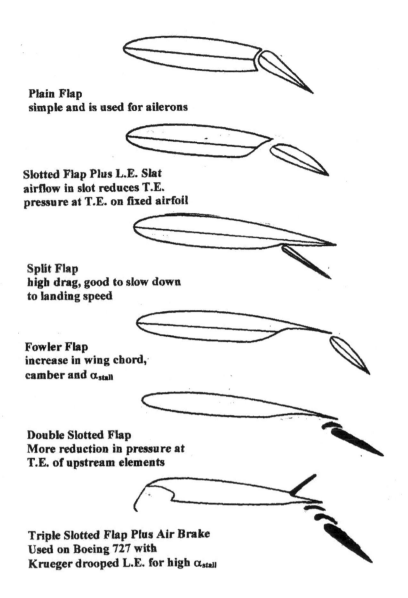

Plain Flap
simple and is used for ailerons

Slotted Flap Plus L.E. Slat
airflow in slot reduces T.E.
pressure at T.E. on fixed airfoil

Split Flap
high drag, good to slow down
to landing speed

Fowler Flap
increase in wing chord,
camber and α_{stall}

Double Slotted Flap
More reduction in pressure at
T.E. of upstream elements

Triple Slotted Flap Plus Air Brake
Used on Boeing 727 with
Krueger drooped L.E. for high α_{stall}

Figure 5-20: Various high lift devices in use capable of increasing maximum lift coefficients to values exceeding 3.

It is also worth noting that the shape and smoothness of the lower surface of a wing has little effect on its performance. This is one of the main reasons that weapon systems and extra fuel tanks are suspended from the bottom of the wing.

For military and rescue operations, short field and navy carrier take-off and landings often require larger lift coefficients than available from the passive high lift devices shown in Figure 5-20. In those cases powered high lift is required. There is a wide range of powered high lift systems. The simple ones use propeller slipstream deflection or fan-jet aircraft slipstream deflection. This technique augments the down wash behind the wing and thus the lift on the wing. Note this lift increase occurs without adding induced drag, since the deflected thrust is mostly fully recovered. Note also the thrust is the deflected jet momentum times the cosine of the deflection angle, which for 30 degrees is 0.87, thus at most a 13% loss in thrust while sin 30 degree is 0.5 or 50% of the thrust is augmenting wing lift. Some techniques are even more efficient, and are shown in Figure 5-21.

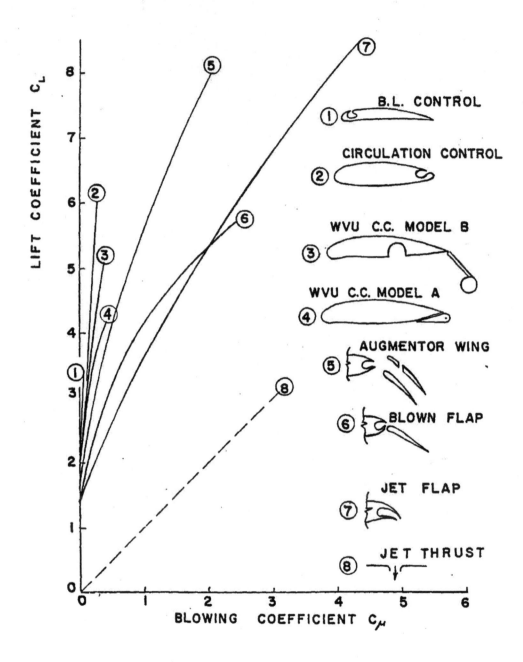

Figure 5-21: Various powered high lift systems. The jet flap can be used for lift coefficients up to 12.

The most power-efficient high lift system is boundary layer control (BLC) by suction. Only a small 5 HP vacuum cleaner is needed to achieve C_{Lmax} up to 3.0 with a single element airfoil. Dr. Raspen conducted flight experiments using BLC at Mississippi State University in the late 50's. His untimely death in a departure stall was blamed on early morning fog which clogged some of the numerous small holes used to remove the boundary layer. Around the same time in England at the University of Cambridge, experiments with BLC in the MK-4 aircraft were underway. The aircraft was capable of stable flight at an unbelievable angle of attack of 45°. Flight at this angle is really not very useful, however, because with such a high nose attitude the pilot cannot see where he or she is going. The program was later terminated when both a pilot and student were killed during a flight experiment.

The second most power efficient technique of powered lift generation is circulation control (CC) by blowing over a rounded trailing edge. Dr. Jones invented this technology at Cambridge University in 1960. In 1964 The Office of Naval Research (ONR) contracted with West Virginia University (WVU) to investigate this technology at full scale flight Reynolds numbers. This resulted in the first Circulation Controlled aircraft, the WVU CC STOL Technology Demonstrator, shown in Figure 5-22. Its first flight was April 15, 1974. In the next 25 hours of flight time all operating characteristics of CC technology were tested and documented, and are reported in "Flight Performance of a Circulation Controlled STOL", by Loth, J.L., Fanucci, J.B. and Roberts, S.C., Journal of Aircraft, Vol. 13, No 3, March 1976. In a power-off glide, this aircraft demonstrated a CC flap lift coefficient of 5.3 and average wing lift coefficient of 4.3. Notice these high C_L values are obtained in a standard flight attitude, so the pilot can see the runway clearly. Another interesting feature is its response to Direct Lift Control (DLC). By opening a dump valve, the CC blowing air can suddenly be stopped. This results in a controlled rapid descent without a nose down pitching moment. This maneuverability is very desirable for Navy carrier STOL landings. To convert the wing back into a streamlined conventional wing, the rounded trailing edge of the Model B wing can be retracted into a slot inside the bottom of the wing, as shown in Figure 5-23.

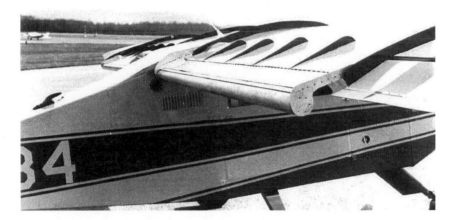

Figure 5-22: The WVU CC STOL Technology Demonstrator (c. 1975).

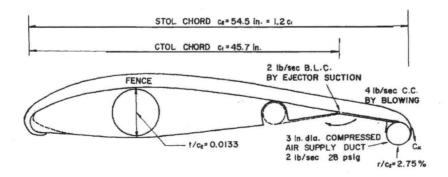

Figure 5-23: The WVU CC STOL Model B wing used a drooped LE and a deployable flap with a rounded TE.

The drooped LE design on the WVU CC STOL wing proved to be very effective as well. In fact, in the 1980's, Cessna equipped all their light aircraft with a similar drooped LE. To estimate the performance of such a CC powered lift wing, one can utilize a simple empirical equation for its performance enhancement. The new variable required is the dimensionless blowing coefficient, C_μ, which relates the jet momentum thrust to the reference force, and is given as

$$C_\mu = \frac{\left(\rho A V\right)_{jet}}{q_\infty S_w} = \frac{\left(\rho t V\right)_{jet}}{q_\infty c} \tag{5.18}$$

The blowing slot height, t_{jet}, is optimum at about 1.5% of the diameter of the rounded trailing edge. The increase in lift coefficient of the system could be calculated using empirically determined relationship:

$$\frac{\Delta C_L}{\Delta C_\mu} = \frac{6.2}{\sqrt{C_\mu}}$$

It was noted that the 3D lift curve slope, a, changed negligibly by CC blowing. The drag coefficient, on the other hand, is reduced by C_μ, as most of the blowing momentum is recovered as thrust. Note CC control is now used in a variety of applications. One interesting application is the elimination of the tail rotor on a helicopter. This is possible by CC blowing over a cylindrical tail boom, where the rotor downwash provides the horizontal lift needed to cancel the rotor shaft torque.

Later in the same decade of the WVU CC STOL research, the US Navy showed interest in CC high lift applications for slow flight in the A-6A attack bomber. This was equipped in 1979 with a similar rounded trailing edge as was on the WVU wing, except that the A-6A CC modification was not retractable. The result was that the cruise speed dropped from 538 knots down to 250 knots. An image of the Navy aircraft is shown in Figure 5-24.

Figure 5-24: The 1979 US Navy A-6A attack bomber used a CC STOL design similar to the WVU CC STOL Technology Demonstrator, but it did not incorporate a retractable TE for efficient high speed cruise.

5.5   COMPUTATIONAL AIRFOIL ANALYSIS

So far, only experimental data has been cited as the primary source of 2D information about wing sections. As with any engineering endeavor, experimental methods are only one way to determine characteristics of a given situation. Two other broad categories also exist: analytical and numerical methods. Analytical methods approach a subject with the intent of discovering a closed-form mathematical solution to a problem. A classic example of this approach in aerodynamics is Paul R. H. Blasius' exact solution method to describe aspects of the laminar boundary layer. For solutions which can only be approximated, it often makes more sense to discretize the governing partial differential equations for a system and in so doing prepare them for various digital computing methods. This approach is commonly referred to as the computational method. This following section will consider how one might be able to compute the lift and drag for a given airfoil shape.

2D POTENTIAL FLOW AND THE VORTEX PANEL METHOD

A common method used to estimate the section lift and drag coefficient from an airfoil shape is to use what is known as the vortex panel method within a potential flow framework. In this approach, the airfoil shape may be composed of any number of flat panels extending between known airfoil coordinates. Each of these panels is stitched together end to end and given varying degrees of vorticity. Each panel's vorticity influences the flow on itself as well as every additional panel. The construct is inserted mathematically into a uniform flow field and placed under the additional constraints that each panel is impermeable (i.e., flow must run tangentially to panels at their surface) and that the pressure and flow direction must be identical at the adjoining upper and lower surfaces of the airfoil TE (aka, the Kutta condition). This approach provides a relatively simple but large system of equations which may be solved rapidly by any modern computer. It is very important to note that the vortex panel method relies on the assumption of inviscid flow, which, as has already been noted, is ideal and not in fact an accurate reflection of reality.

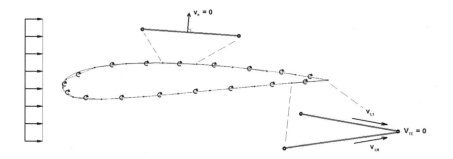

2D INTEGRAL BOUNDARY LAYER METHOD AND TRANSITION AND SEPARATION MODELING

Chapter 4 discussed how well-established behavioral trends for laminar and turbulent boundary layers on flat plates may be used to compute the viscous drag exerted on them – recall Eq. (4.25) and Eq. (4.26), for example. It stands to reason that this approach may also be used to enhance the vortex panel method for airfoil analysis, especially because the flow is essentially localized flat plate flow. If there is some way of knowing when a boundary layer "should" transition from laminar to turbulent, then it is possible to determine the local viscous effects of the fluid on each panel. Furthermore, if it is possible to predict when a turbulent boundary layer "should" separate, then other empirical models for viscous drag may be utilized over stalled panels. By integrating the viscous effects over the entire panelized construct, one may begin to estimate with relatively good accuracy the section lift, drag, and moment generated by the virtual flow.

JAVAFOIL AND XFLR-5

Two very user friendly and absolutely free enhanced vortex panel method pieces of software exist for even the casual airfoil designers: Dr. Martin Hepperle's JavaFoil, and Dr. Mark Drela's Xfoil. JavaFoil is available to use as a web app, but can also be downloaded and run offline as long as the user has some version of Java installed. The JavaFoil download may be found at:

http://www.mh-aerotools.de/airfoils/java/javafoilinstaller.msi

Xfoil, which was designed to interface with several compilers such as MATLAB®, can be conveniently run in its own GUI by downloading the shell program, XFLR-5. This software may be freely downloaded following the "Downloads" link at:

http://www.xflr5.com/xflr5.htm

Whichever software the user may choose to use, the processing must begin with the geometry of the airfoil. Many sources are available for airfoil coordinates. One really exceptional airfoil coordinate website is maintained by Dr. Michael Selig at UIUC:

http://aerospace.illinois.edu/m-selig/ads/coord_database.html

It is also possible to locate some experimental data for a select group of airfoils at the same basic website:

http://aerospace.illinois.edu/m-selig/pd.html

From this location, the user will need to save a copy of the archives by selecting one of the three "[zip archive]" buttons to access the lift and drag data in root directories for text files.

EXAMPLE 5.4: COMPUTATIONAL ANALYSIS OF A WING SECTION

For this example, select some arbitrary airfoil to evaluate. A helpful way to do this is to pick one that looks interesting from the available .gifs on the UIUC coordinate database. Use both JavaFoil and XFLR-5 to get results for a given Reynolds number that also has known experimental values from the NACA data, the UIUC LSATs database, or any other similar data source. For this particular example, select the Verbitsky BE50 Free Flight Airfoil, which is shown in Figure 5-25.

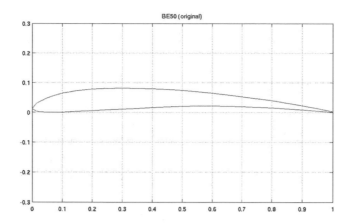

Figure 5-25: The Verbitsky BE 50 airfoil (UIUC .gif).

From the coordinates provided on UIUC's airfoil coordinate database, save the BE50.dat file to the computer.

Use JavaFoil first by opening the software and uploading the coordinates by using the "Open…" button on the "Geometry" tab. Navigate to the "Polar" tab and input the Reynolds number value as 61,000 (since that is fairly close to the value at which it was tested at UIUC). Click on the "Analyze It!" button. The user may then copy the data to a text file by using the "Copy (Text)" button and may also print the output to a Word or PDF file. The output will look like that shown in Figure 5-26.

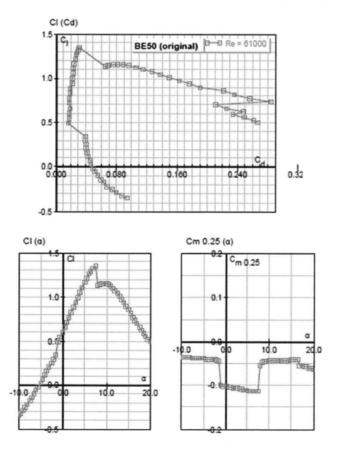

Figure 5-26: BE 50 2D data at a Reynold number of 61,000 as computed by JavaFoil.

Next, use XFLR-5 to evaluate the same airfoil. Open the software and start the process by clicking on the "File" option at the upper left corner. Select "XFoil Direct Analysis" from the tree. Next, click on the blue folder icon to the left of the disk icon. Now navigate to the BE50.dat file and click on it. Now with the coordinates loaded onto the page, select "Define an Analysis" in the tree under the "Analysis" button. Input the operating Reynolds number of 61,000 and click "OK". Now on the right hand side of the view window, note there is a range of angles of attack and degree increments. After entering the values desired, click the "Analyze" button. Export the data by clicking on "Polars", then "Current Polar", then "Export" to write the data in a text file. The output is shown in Figure 5-27.

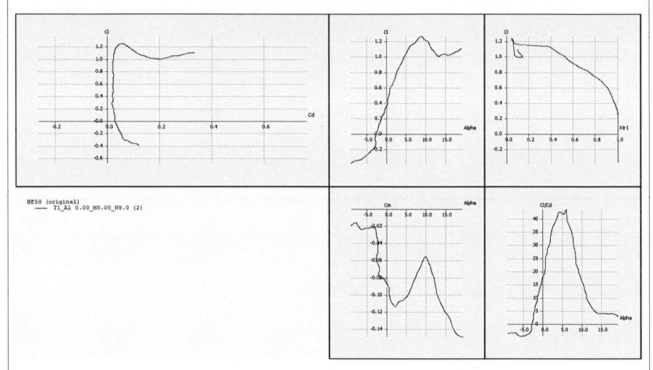

Figure 5-27: BE 50 2D data at a Reynold number of 61,000 as computed by XFLR-5.

Taking the computational data and comparing it to the experimental data collected at UIUC, plots of the results may be generated in MS Excel®, as shown in Figure 5-28 and Figure 5-29.

From the resulting graphs, it is evident that there is pretty good agreement between the computational and experimental results. There are some significant differences for some key characteristics, however. It is evident from the C_l versus α curves that JavaFoil's results for the zero lift angle of attack, maximum section lift, and associated stall angle are not quite right when compared to the experimental results. XFoil's results tend to over-predict the lift curve slope value. Drag results tend to be in very good agreement.

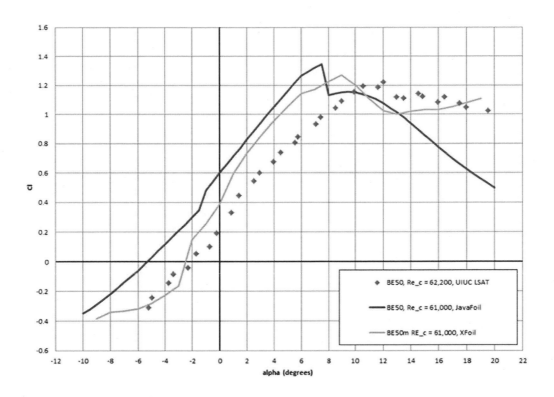

Figure 5-28: Section lift comparison between UIUC experimental, and JavaFoil and XFoil numerical results.

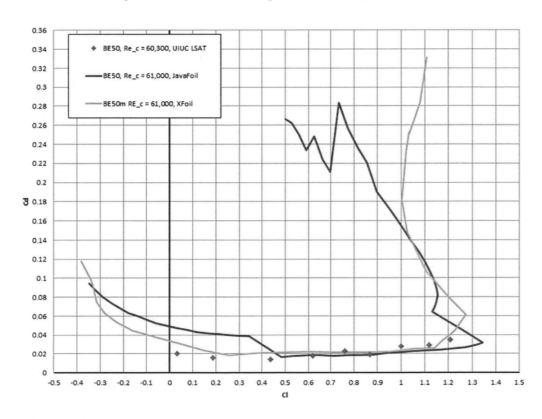

Figure 5-29: Section drag comparison between UIUC experimental, and JavaFoil and XFoil numerical results.

VORTEX LATTICE METHOD FOR 3D LIFTING SURFACES

There is a 3D analog of the vortex panel method which may be applied in similar, though not identical, ways to provide insight into lift and drag being generated by 3D bodies in a flow. The technique itself, known as the vortex lattice method, was first formulated in the late 1930's, but was not really used effectively until the advent of the modern digital computer during the 1960's. Of particular interest to the motivated aerospace engineering student is the relatively easy access that XFLR-5 gives to utilizing this numerical approach for wings and even for entire gliders. It is left to the eager student to do the necessary investigation into the use of such a program, but the results can be quite satisfying when trying to design a glider from scratch.

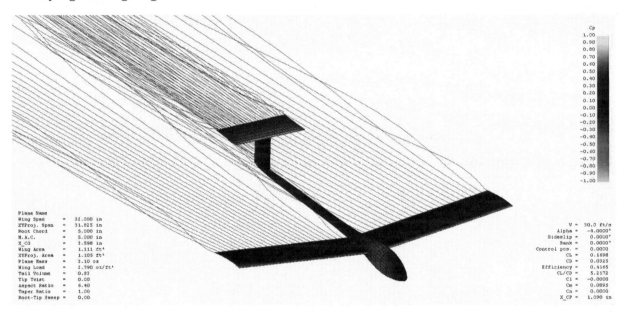

Figure 5-30: Isometric view of a conceptual glider with pressure coefficients and flow streamlines shown.

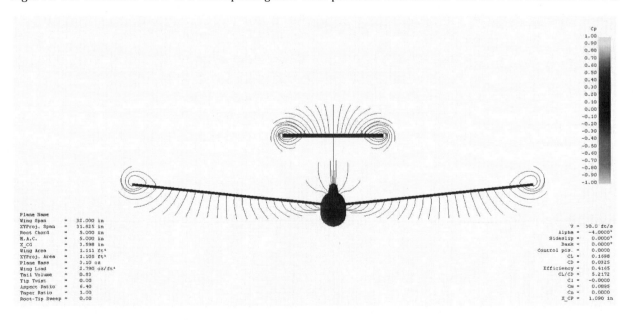

Figure 5-31: Front view of a conceptual glider with pressure coefficients and flow streamlines shown.

5.6 SUMMARY

ANATOMY OF THE AIRFOIL

In general, students should be able to identify, describe, and/or sketch the following items on a wing section:

1. Mean line – this is the locus of points for which a straight line drawn normal to its local path and extending from upper surface to lower surface is bisected
2. Chord line – this is the straight line path between the LE and TE
3. Thickness – this is the maximum distance between the upper and lower surface, measured perpendicular to the chord line (usually stated in terms of per cent chord)
4. Chord length – this is the length of the chord line (i.e., the distance between the LE and TE)
5. Upper & lower surfaces – the surface that extends from the LE to the TE along the upper side of the airfoil is the upper surface; the surface that extends from the LE to the TE along the lower side of the airfoil is the lower surface
6. LE & TE – the LE is the forward-most part of the airfoil when positioned at a 0° angle of attack; the TE is the aft-most part of the airfoil when positioned at a 0° angle of attack
7. Quarter-chord location – this is the location 25% of the chord length aft of the LE along the chord line

DYNAMICS OF 2D AIRFOILS

A wing section derives its lift from the bound circulation around the shape. The circulation is quantified as vortex strength, Γ, where

$$\Gamma = \oint \vec{V} \cdot d\bar{s}$$

And the lift per unit span is calculates as

$$L' = \rho_\infty V_\infty \Gamma$$

2D AIRFOIL TEST DATA

This section primarily covers the following items:

1. Definitions of section lift, drag, and moment coefficients:

$$C_l = \frac{L'}{q_\infty c}$$

$$C_d = \frac{D'}{q_\infty c}$$

$$C_{m,c/4} = \frac{M'_{c/4}}{q_\infty c^2}$$

2. Sign convention for section lift, drag, and moment:

$$\text{Lift up} \qquad\qquad L'(+)$$
$$\text{Drag aligned } w / V_\infty \qquad D'(+)$$
$$\text{Pitching moment nose} - up \qquad M'_{c/4}(+)$$

3. Read the following items from a traditional (NACA-type) 2D data graph:

   a. $C_{l,max}$ – this is maximum value of section lift coefficient
   b. α_{stall} – this the value of angle of attack corresponding to the maximum section lift coefficient
   c. $\alpha_{L=0}$ – this is the value of angle of attack where the section lift coefficient is equal to zero
   d. $C_{d,0}$ – this is the minimum value of section drag coefficient

4. Calculate the following from 2D data:

$$a_0 = \frac{\Delta C_l}{\Delta \alpha}$$
$$k = C_{d,C_l=1.0} - C_{d,0}$$
$$C_l(\alpha) = a_0 (\alpha - \alpha_{L=0})$$
$$C_d(C_l) = C_{d,0} + kC_l^2$$

NACA AIRFOILS

NACA four-digit airfoil numbering scheme:

- First digit: The maximum camber of the airfoil in per cent chord length
- Second digit: The location of the maximum camber, measured from the LE, in 10 percent chord length
- Last two digits: The thickness of the airfoil in per cent chord length

Example: The NACA 4412 airfoil has a maximum camber of 4% chord length which occurs at a location 40% of the chord length aft of the LE. The airfoil is 12% of the chord length in thickness.

LIFT INDUCED DRAG, ASPECT RATIO, AND SPANWISE LOADING

This section covered the phenomenon of induced drag due to the rotation of the lift vector aft as a result of finite wing downwash. It is important to understand that aspect ratio, AR, plays a pivotal role in determining the severity of wing downwash and lift-induced drag.

$$AR = \frac{b^2}{S_w}$$

For a rectangular wing, aspect ratio reduces to the ratio of the span to the chord length such that

$$AR_{rect} = \frac{b}{c}$$

Oswald's efficiency, *e*, gives an indication of loading efficiency. Elliptical loading is *e* = 1.00. Empirically, *e* is correlated to the aspect ratio such that

$$e = \frac{2}{\left[2 - AR + \sqrt{\left(4 + AR^2 \right)} \right]}$$

Wing planform, twist, and taper may be used to maximize loading efficiency by minimizing the pressure gradient at the wingtips.

2D AIRFOIL TO 3D WING PERFORMANCE CONVERSIONS

To convert 2D to 3D lift curve slopes,

$$a = \frac{a_0}{\left(1 + \dfrac{57.3\, a_0}{\pi e AR} \right)}$$

From which the 3D lift coefficient may be obtained as

$$C_L = a \left(\alpha - \alpha_{L=0} \right)$$

To calculate 3D wing lift, drag and pitching moment,

$$C_L = \frac{L}{q_\infty S_w}$$

$$C_D = \frac{D}{q_\infty S_w}$$

$$C_{m,c/4} = \frac{M_{c/4}}{q_\infty S_w c}$$

For subsonic flows, to calculate total drag for a finite wing, one must include the lift-induced drag so that

$$D_{total} = D_{para} + D_i$$

Expanding this relationship,

$$D_{total} = \left(C_{D,para} + C_{D,i} \right) q_\infty S_w$$

Or , finally,

$$D_{total} = \left(C_{d,0} + \frac{C_L^2}{\pi e AR} \right) q_\infty S_w$$

WING CRUISE VELOCITY

The key for cruise condition lies in the maximization of aircraft lift-to-drag (L/D). Since lift is typically equivalent to the aircraft weight, the only way to maximize L/D is to minimize drag. This is accomplished by ensuring equivalent parasitic and induced drag which gives and optimum lift coefficient,

$$C_{L,opt} = \sqrt{\left(C_{D,para} \pi e AR \right)} = \frac{W}{\frac{1}{2} \rho V_{cruise}^2 S_w}$$

The associated cruise velocity may then be calculated as

$$V_{cruise} = \sqrt{\frac{2W}{\rho C_{L,opt} S_w}}$$

WING STALL VELOCITY

The limit to the amount of lift a wing can generate is directly connected to the maximum lift coefficient attainable by the wing section upon which the wing is based. Generally speaking, the maximum *wing section* lift coefficient is the same as the maximum *wing* lift coefficient. The stall velocity may be found by first assuming that the lift is equivalent to the aircraft weight and then solving for the minimum dynamic pressure required to maintain lift at the maximum lift coefficient, or

$$q_{stall} = \frac{W}{C_{L,max} S_w}$$

Once the minimum dynamic pressure is known, then the stall velocity is readily calculated as

$$V_{stall} = \sqrt{\frac{2W}{C_{L,max} \rho S_w}}$$

CHAPTER 5 PROBLEMS

5.1) What is the difference between a symmetric and a cambered airfoil in terms of the chord and camber lines?

5.2) Make a sketch of a cambered airfoil at positive angle of attack. Label all the critical components of the airfoil and include the freestream velocity vector and angle of attack.

5.3) Create a sketch of the lift curve slope *and* the drag polar for a symmetric airfoil with the following performance data:

$$a_0 = 0.105 \left(1/^\circ\right) \qquad C_{l,\max} = 1.15 \quad \alpha_{stall} = 14^\circ \qquad C_{d,0} = 0.005 \qquad k = 0.002$$

Be sure to label all axes and indicate numerical values for scales.

5.4) Using Figure C-4 from Appendix C: Sample Airfoil Data to find/calculate the following values:

    a. Find $C_{L,max}$ for Re_c = 3.0 x 10^6
    b. Find $\alpha_{L=0}$ for Re_c = 9.0 x 10^6
    c. Find $C_{d,0}$ for Re_c = 6.0 x 10^6
    d. Find k for Re_c = 3.0 x 10^6

5.5) Using Figure C-1 from Appendix C: Sample Airfoil Data to find/calculate the following values:

    a. Find $C_{m,c/4}$ for Re_c = 3.0 x 10^6
    b. Find $\alpha_{L=0}$ for Re_c = 9.0 x 10^6
    c. Find $C_d(C_l=0.7)$ for Re_c = 3.0 x 10^6
    d. Find a_0 for Re_c = 6.0 x 10^6

5.6) A 2D experimental test is run under the following conditions:

$$V_\infty = 15.7 \, m/sec \qquad \rho_\infty = 1.123 \, kg/m^3 \qquad c = 4 \, in \qquad b = 18 \, in \qquad C_l = 0.615$$

Calculate the following values:
    a. Find the tunnel dynamic pressure, q_∞
    b. Find the aspect ratio, AR, of the wing
    c. Find the lift per unit span, L', of the wing
    d. Determine the vortex strength, Γ, of the bound circulation on the wing at that condition

5.7) Consider the vortex strength at any arbitrary downstream x-location in the boundary layer of a flat plate at zero degree angle of attack. Defining the control volume as a rectangular form with unit length having its base on the plate's surface with a height equal to the local thickness of the boundary layer, δ, show that the vortex strength is constant and is equal to U_∞ m²/sec.

5.8) List the advantages of twist and taper on a 3D wing.

5.9) What advantage does flying in a V-formation give to migrating birds? Draw a sketch to help support your explanation.

5.10) An untwisted, tapered wing moves through standard sea level air under the following conditions:

$$c_r = 2.0\,m \qquad c_t = 0.5\,m \qquad b = 8\,m \qquad C_{d,0} = 0.0045 \qquad V_\infty = 50\,m/sec$$

Calculate the following values:
- a. Find the wing's taper ratio, λ
- b. Find the wing's aspect ratio, AR
- c. Estimate the Oswald's efficiency, e, of the wing
- d. Calculate the optimum lift coefficient, $C_{L,opt}$, for the wing
- e. Find the mean aerodynamic chord-based Reynolds number, $Re_{M.A.C.}$, of the wing

5.11) A rectangular wing is based upon an airfoil with the following characteristics:

$$2D\,Data: \qquad a_0 = 0.100\,(1/^\circ) \qquad \alpha_{L=0} = -2^\circ \qquad C_{d,0} = 0.003$$

$$3D\,Data: \qquad \alpha = 6^\circ \qquad V_\infty = 22\,m/sec \qquad \rho_\infty = 1.109\,kg/m^3$$

$$e = 0.83 \qquad b = 8\,m \qquad S_w = 6\,m^2$$

Calculate the following values:
- a. Find the value of the lift, L, generated by the wing
- b. Find the value of the drag, D, generated by the wing

5.12) A Cessna Skyhawk 172S has the following design specifications:

$$W = 2{,}550\,lb_f \qquad C_{L,max} = 1.215 \qquad S_w = 174\,ft^2 \qquad V_{cruise} = 143\,mph \qquad b = 36\,ft$$

Calculate the following values:
- a. Find the stall velocity, V_{stall}, of the Skyhawk 172S
- b. Find the value of the optimum lift coefficient, $C_{L,opt}$, of the Skyhawk 172S

5.13) Recreate the MS Excel® Spreadsheet as shown below, completing all the values of V_∞ between 90 and 320 ft/sec in increments of 5 ft/sec, then plot drag as a function of V_∞, labeling both axes.

	A	B	C	D	E	F	G	H
1	2D Airfoil Data					2D Wing Test Specifications		
2	$\alpha_{L=0}$	(°)	-2			b	(ft)	30
3	a_0	(1/°)	0.095			c	(ft)	5
4	$C_{l,max}$	(~)	1.25			S_w	(ft²)	150
5	$C_{m,c/4}$	(~)	-0.05			L	(lb$_f$)	1500
6	$C_{d,0}$	(~)	0.008					
7	k	(~)	0.005			Operating Conditions		
8						ρ_∞	(slug/ft³)	0.002377
9								
10	V	q	C_l	α	C_d	C_l/C_d	D	$M_{c/4}$
11	(ft/sec)	(lb$_f$/ft²)	(~)	(°)	(~)	(~)	(lb$_f$)	(ft-lb$_f$)
12		=0.5*$H$8*A12^2	=$H$5/(B12*$H$4)	=C12/$C$3+$C$2	=$C$6+$C$7*C12^2	=C12/E12	=E12*B12*$H$4	=$C$5*B12*$H$4*$I$3
13	85	8.59	1.165	10.3	0.0148	78.8	19.0	-322
14	90	9.63	1.039	8.9	0.0134	77.5	19.3	-361
60	320	121.70	0.082	-1.1	0.0080	10.2	146.7	-4,564
61	325	125.54	0.080	-1.2	0.0080	9.9	151.2	-4,708

CHAPTER 6.   AIRCRAFT PERFORMANCE ANALYSIS

6.1   INTRODUCTION

Aircraft are designed and built for numerous reasons, but by and large they are used for relatively high-speed transport of passengers, cargo, or both. This chapter will focus on several critical aspects of a modern aircraft designed to carry both passengers and cargo, including

- Aircraft attitude – accepted norm for orientation of the aircraft with respect to the horizon
- Propulsion & power –forward motive force required for better-than-glide flight
- Banking – maneuvering required to make turns in flight
- Endurance – indicates how long the aircraft can remain aloft
- Range – indicates how far the aircraft can travel
- Takeoff and landing distances – critical to establish the runway lengths required for safe departure and arrival at airports
- Service ceilings – indicates how high the aircraft can possibly fly

Before going into the details of these performance specifications, a brief background into the guiding constraints for modern passenger and transport aircraft is necessary to gain a full appreciation for the delicate balances that must be made in the overall design of an air-worthy plane.

SPEED AND ALTITUDE: IMPORTANT DESIGN CONSTRAINTS FOR MODERN AIRCRAFT

The economy of an aircraft is perhaps the first and arguably the most important design constraint: greater distance per ton of cargo or per passenger means greater economic viability inherent in the design. Simply considering the previous statement implies that one way to improve the passenger miles-per-gallon may be to increase the speed of the aircraft. The reader is asked to recall the lessons learned in Section 5.3, particularly the determination of an explicit cruise velocity at which the maximum lift-to-drag ratio is achieved (see Figure 5-19). Without question, operation at maximum lift-to-drag (also referred to sometimes as glide ratio or finesse) always implies maximum fuel economy. However, given that drag coefficients appear to be sensitive to Reynolds number only at low speeds – recall Figure 4-33 and Figure 4-34 where the drag coefficients appear to stabilize at higher Reynold numbers – is it not possible to increase the cruise velocity by several factors by simply reducing the wing area of the aircraft? It turns out that this is true to a certain extent, but one eventually runs into a problem when the flow around the aircraft becomes highly compressible. Even at Mach numbers much less than 1.0, the speed of air moving around the aircraft can reach sonic and supersonic speeds, with resultant shock waves forming on or near the aircraft's surface. Shock waves are incredibly productive entropy generators and are thus responsible for massive losses on flow total pressure near the aircraft resulting in dramatic increases in drag through the introduction of a new drag component known as wave drag. The sudden rise in drag coefficient at these transonic speeds gives rise to the term "sound barrier". The use of the word "barrier" is indicative of early designers' uncertainty about the ability for manned aircraft to go through such a flight regime without dire consequences to the pilot and the aircraft. The sound barrier was officially broken on October 14, 1947 by West Virginia native Army Air Force Captain Charles "Chuck" Yeager flying in the Bell X-1 *Glamorous Glennis* at speeds up to Mach 1.06. NACA, Army Air Force, and Bell engineers had designed the X-1 fuselage like a bullet based on the known supersonic stability of the Browning

.50 Caliber Machine Gun bullet. Bullets had long been known to move faster than the speed of sound, thus the sound barrier was never thought of by any serious scientist or engineer as an absolutely impenetrable flight speed. Results of standard US Army drag tests of the NATO 7.62mm M80 bullet are presented in Figure 6-1 to help illustrate the significant increase in drag near sonic speeds.

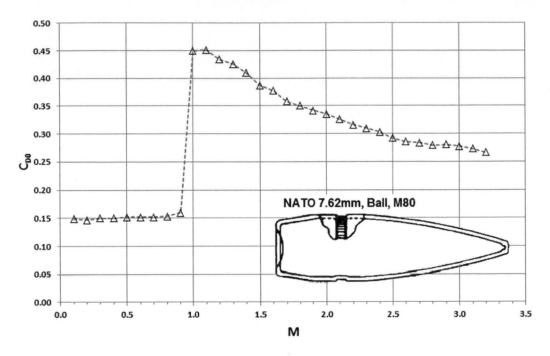

Figure 6-1: Minimum drag coefficient of standard ammunition shown at varying flight Mach numbers.

Considering the sound barrier with respect to passenger and transport aircraft, an important conclusion is that to fly economically, the aircraft must fly as fast as possible within the subsonic regime. Any slower and ton-miles or passenger-miles per gallon are lost; any faster and maximum operational lift-to-drag ratios are drastically reduced as a result of the sudden onset of wave drag.

Another important constraint for modern aircraft is based upon operational altitude. Depending upon the powerplant and propulsion system onboard, the impact of the local air temperature and density is extremely significant. Nearly all aircraft propulsion systems rely on the combustion of hydrocarbon fuels. Avgas is used for spark-ignited internal combustion (IC) engines and jet fuel is used for gas-turbine engines. The combustion process burns air mixed with fuel which must be maintained at the proper mixture ratio for stable and efficient engine operation. Spark-ignited IC engines that are typical on small propeller-driven light aircraft suffer at higher altitudes where the density of the ambient air, and thus the oxygen content per volume of intake air, is severely reduced in comparison to sea level values. On the other hand, as discovered by the French engineer Nicolas Carnot, higher thermal efficiency of heat engine processes is dependent only upon the temperatures of the two "reservoirs" between which the engine operates. The famed Carnot cycle operates with the maximum permissible thermal efficiency allowable and is found as

$$\eta_{Carnot} = 1 - \frac{T_C}{T_H} \qquad (6.1)$$

where T_C and T_H are the absolute temperature values of the cold and hot reservoirs, respectively. In the case of aircraft engines, T_C is the temperature of the ambient air and T_H is the maximum temperature reached during the combustion process in the engine. Carnot efficiency implies that an engine must operate where the air temperature is colder to achieve its best performance. Gas-turbine engines, especially the modern high bypass ratio turbojet, achieve remarkable efficiencies operating at high altitudes, but in order to maintain thrust they must "breathe" as much air into their intake as possible. Moving at high subsonic speeds has an additionally beneficial result of pre-compressing the intake air in front of the engine to increase the efficiency even more. Although prop-driven aircraft typically do not fly as high as jet aircraft, both take advantage of the lower density of the ambient air at higher altitudes in one other respect: since the weight, lift coefficient, and wing area are fixed for a given plane in level flight and since the density is lower at higher altitudes, the flight speed must be increased to maintain adequate dynamic pressure for lift equal to weight. The benefit of higher flight altitude is thus a faster traveling speed, which again reinforces the overall economy of flight as a means of transport and travel.

A WORLD OF FLYERS

While manned flight may still be considered news, flight in and of itself has existed as a biological phenomenon for ages. Birds, bats, insects, and even plants and plant seeds routinely take to the air as a primary survival capability. In his excellent book The *Simple Science of Flight: From Insects to Jumbo Jets*, author Hendrik (Henk) Tennekes showcases and expounds on what is now commonly referred to as The Great Flight Diagram. Though at first glance there is a world of difference between a Boeing 747 and the common fruit fly, *Drosophila melanogaster*, there are amazingly similar relationships common to both designs. In particular, a relationship appears to exist between the wing area, weight, and cruise speed of most flyers so that the wing loading can be expressed as

$$W/S_w \approx 0.38 V_{cruise}^2 \qquad (6.2)$$

and

$$W/S_w \approx 47 W^{\frac{1}{3}} \qquad (6.3)$$

Eq. (6.2) stems from the fact that all flyers need to generate lift equal to their own weight: a simple rearrangement of the terms will merely confirm that the weight of the aircraft in level flight is the product of the lift coefficient, dynamic pressure, and wing area. Eq. (6.3) helps underscore the fact that weight goes as the cube of characteristic length. Taking the span, *b*, as the characteristic length, it is easy to see that the weight is proportional to *b³* and the span is of course proportional *b²*, thus the wing loading is proportional to *b*, which is itself proportional to the cubed root of the weight. The constants given in both equations should be understood as being indicative of a general trend, rather than taken as a hard and fast rule to aircraft design. Designs that require extreme operational conditions may depart significantly from the general trends to achieve their specific goals and objectives.

Flyers and their associated operational specifications are shown in Table 6-1, and relationships between critical operating parameters are presented graphically in Figure 6-2 through Figure 6-4. Of particular interest is the slight variation between Eq. (6.2) and Eq. (6.3) when compared to the curve-fit functions shown in Figure 6-2 and Figure 6-3.

Table 6-1: A very brief listing of some of the world's flyers and their associated flight specifications.

Name of flyer	W (N)	S_w (m^2)	b (m)	AR (~)	W/S_w (Pa)	V_{cruise} (m/sec)	$h_{G,cruise}$ (ft)	T_{cruise} (K)	ρ_{cruise} (kg/m^3)	$C_{L,opt}$ (~)
Airbus A 380-800	4100580	845	79.8	7.53	4852.8	252.1	35000	218.93	0.3802	0.402
Boeing 747-400	3901705	560	59.7	6.37	6967.3	253.6	35000	218.93	0.3802	0.570
Lockheed Martin SR-71	891818	167	16.9	1.72	5347.9	944.2	85000	216.66	0.0352	0.341
McDD DC 9-80	665118	118	32.9	9.18	5636.6	225.3	35000	218.93	0.3802	0.584
Boeing 737-200	514000	91	28.9	9.18	5648.4	216.8	35000	218.93	0.3802	0.632
Airbus A 320-200	510120	123	35.8	10.45	4160.8	230.0	36000	216.96	0.3659	0.430
Bombardier CRJ-700	245250	69	23.2	7.85	3575.1	230.6	35000	218.93	0.3802	0.354
Saab 2000	186390	56	24.8	11.04	3346.3	155.6	35000	218.93	0.3802	0.727
Northrup Grumman RQ-4 Global Hawk	117720	50	35.4	25.04	2349.7	159.7	65000	216.66	0.0915	2.015
Cessna Citation CJI	47149	22	14.3	9.26	2143.1	108.1	41000	216.66	0.2883	1.273
Cessna 206 (WVU-66)	16013	16	11.2	7.77	1000.8	72.9	10000	268.36	0.9047	0.416
Piper Archer III ('98)	11342	16	10.8	7.32	708.9	65.8	12000	264.40	0.8493	0.386
Piper Cherokee	8829	15	9.2	5.59	583.2	55.6	12000	264.40	0.8493	0.445
General Atomics MQ-1B Predator	8829	12	16.8	24.54	767.7	39.6	22000	244.62	0.6099	1.606
German glider ETA	8339	19	31.0	51.67	448.3	22.2	10000	268.36	0.9047	2.007
Cessna 150 WVU-'59	6672	15	10.2	6.94	444.8	55.0	10000	268.36	0.9047	0.325
AeroVironment Gossamer Albatross	956	45	29.8	19.55	21.1	6.9	500	287.17	1.2073	0.724
Mute swan	106	0.650	2.30	8.14	163.1	16.2	500	287.17	1.2073	1.029
Whooper swan	86.9	0.605	1.98	6.48	143.6	18.5	500	287.17	1.2073	0.695
Tundra swan	66.4	0.461	1.98	8.50	144.0	18.5	500	287.17	1.2073	0.697
Common crane	56.1	0.586	2.22	8.41	95.7	15.0	500	287.17	1.2073	0.705
Golden eagle	40.7	0.597	2.03	6.90	68.2	11.9	500	287.17	1.2073	0.797
Canada goose	36.3	0.372	1.69	7.68	97.6	16.7	500	287.17	1.2073	0.580
White stork	34.0	0.500	2.00	8.00	68.0	13.0	500	287.17	1.2073	0.667
Great white heron	25.0	0.490	1.91	7.45	51.0	11.1	500	287.17	1.2073	0.686
Blue heron	19.2	0.420	1.76	7.38	45.7	11.1	500	287.17	1.2073	0.615
Osprey	15.8	0.320	1.60	8.00	49.4	13.3	0	288.16	1.2252	0.456
Turkey vulture	15.5	0.440	1.75	6.96	35.2	8.0	1000	286.18	1.1897	0.925
Pheasant	12.0	0.088	0.72	5.89	136.4	18.6	0	288.16	1.2252	0.643
Raven	11.5	0.247	1.21	5.93	46.6	14.3	0	288.16	1.2252	0.372
Herring gull	11.4	0.197	1.34	9.11	58.0	12.6	0	288.16	1.2252	0.596
Red tailed hawk	10.9	0.209	1.22	7.12	52.2	9.8	0	288.16	1.2252	0.886
Peregrine falcon	7.89	0.126	1.02	8.28	62.8	12.1	0	288.16	1.2252	0.700
Ruffed grouse	5.17	0.053	0.56	5.92	97.5	15.6	0	288.16	1.2252	0.654
Barn owl	5.05	0.168	1.12	7.47	30.1	13.4	0	288.16	1.2252	0.273
Hen harrier	4.33	0.157	1.10	8.50	27.6	9.1	0	288.16	1.2252	0.544
Sparrow hawk	2.50	0.080	0.75	7.03	31.3	9.9	0	288.16	1.2252	0.520
Common tern	1.20	0.056	0.83	12.30	21.4	13.6	0	288.16	1.2252	0.189
Blue jay	0.89	0.024	0.38	6.02	37.1	10.0	0	288.16	1.2252	0.605
Robin	0.82	0.024	0.38	6.02	34.2	10.9	0	288.16	1.2252	0.469
House sparrow	0.28	0.01	0.23	5.29	28.0	12.5	0	288.16	1.2252	0.293
Ruby throated hummingbird	0.03	0.0012	0.09	6.75	25.0	11.1	0	288.16	1.2252	0.331
Dragonfly	0.01	0.0018	0.10	5.56	5.6	4.5	0	288.16	1.2252	0.448
Swallowtail butterfly	0.003	0.006	0.08	1.07	0.5	2.2	0	288.16	1.2252	0.169
Crane fly	0.0003	0.00008	0.02	5.00	3.8	3.0	0	288.16	1.2252	0.680
Fruit fly	0.000007	0.000003	0.003	3.00	2.3	2.0	0	288.16	1.2252	0.952

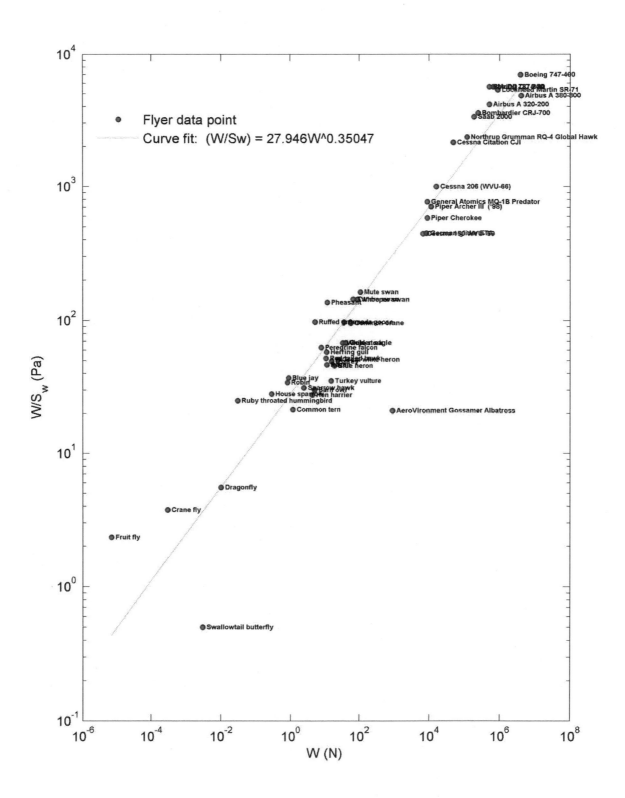

Figure 6-2: The Great Flight Diagram (modified from Tennekes).

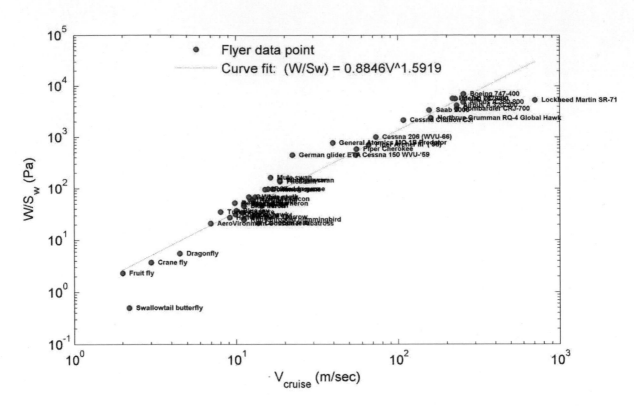

Figure 6-3: Variation of the Great Flight Diagram showing the relationship between wing loading and cruise speed.

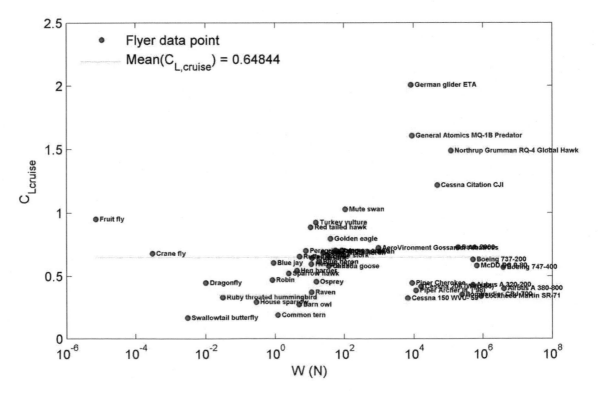

Figure 6-4: Although somewhat diverse, the average optimum lift coefficient for most flyers is about 0.649.

6.2   AIRCRAFT ATTITUDE

To ensure engineers and designers are speaking about the same thing when it comes to flight maneuvers and sides of an aircraft, it is important to establish a "correct" view of the aircraft in terms of attitude convention. Aircraft attitude is simply a way of orienting the aircraft with respect to the flight horizon. In flight, an aircraft has freedom of motion to translate and rotate about three primary axes:

- The longitudinal axis, also known as the x-axis or roll axis, allows for forward and back translation as well as rolling rotation. From the pilot's perspective, the convention is that forward translation and clockwise rotation are positive.
- The lateral axis, also known as the y-axis or pitch axis, allows for port and starboard translation as well as pitching rotation. From the pilot's perspective, the convention is that starboard (rightward) translation and pitch-up rotation are positive.
- The vertical axis, also known as the z-axis or yaw axis, allows for ascent and descent translation as well as yawing rotation. From the pilot's perspective, the convention is that descent (downward) translation and clockwise rotation are positive.

Figure 6-5 shows a typical aircraft along with the primary axes and direction sign conventions.

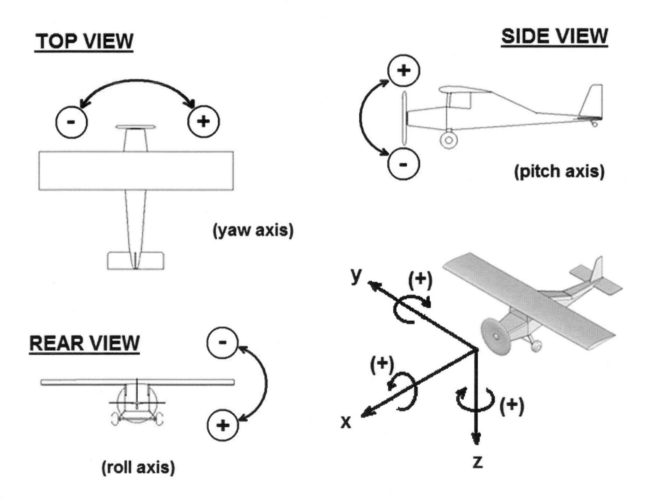

Figure 6-5: Accepted aircraft attitude conventions.

6.3   CRUISE VELOCITY VERSUS WING LOADING AND ALTITUDE

As discussed in this chapter's introduction, the operational altitude of an aircraft depends on several important factors. For any aircraft it is desirable to fly at the constant optimum lift coefficient which is usually between 0.6 and 0.7 to maintain peak glide ratio and flying efficiency. The cruise velocity at which the optimum glide ratio occurs varies, but all aircraft must fly at their maximum efficient speed. In the case of a commercial aircraft, for example, the desired speed approaches sonic speed and is typically between Mach 0.8 and 0.85. The combination of aircraft weight, wing area, designed wing loading, identified optimum cruise lift coefficient, and maximum efficient speed occurs when the local air density is much lower than at sea level (usually at an altitude of around 30,000 ft.).

An aircraft's target operational condition is based primarily on its designed wing loading (see Figure 6-2 and Figure 6-3 to get a better feel for the importance of wing loading in various flyers). Wing loading is often one of the first things considered in the design of a new aircraft as it has dramatic effects on every other aspect of the vehicle design. Aerospace engineers must take into account the desired flight speed, take-off requirements, expected aerodynamic coefficients, and selected powerplants and propulsion systems when considering the overall wing loading design. One thing that has not been previously mentioned, however, is that since the aircraft is burning fuel at rates that can exceed 3,500 US gallons (about 21,000 pounds) per hour, the aircraft wing loading will constantly change if kept at a constant altitude and velocity. Usually commercial aircraft start cruise at 30,000 feet where the density is about three times lower than at sea level. Then the optimum cruise speed becomes about 1.73 times greater than at sea level. As the tanks continue to empty, the wing loading reduces which means the aircraft needs to climb steadily to reach about 40,000 ft. where the air density is four times lower than at sea level. At that altitude, the optimum cruise speed doubles that at sea level.

Throughout the variable wing loading conditions and changing flight altitudes, the important thing is that C_L remains the same and is kept at its optimum value so that the aircraft lift-to-drag ratio will also remain at its maximum value. It is interesting to note that with increasing altitude the optimum velocity increases and thus also the required thrust power. Jet-propelled aircraft, whose engines produce more or less constant thrust independent of flight speed, find it advantageous to fly as fast as possible to minimize trip time, whereas prop-driven aircraft with IC engines suffer significant losses in available power at higher altitudes and are thus forced to fly "low and slow".

6.4   PROPULSION AND POWER

All aircraft must utilize some type of propulsion system if it is required to fly at anything better than its maximum descent glide ratio. In fact, no aircraft can even maintain steady level flight without some thrust to overcome inevitable drag. Several different types of powerplants and propulsion systems exist for aircraft, including electric motor-driven systems, gas-electric hybrid systems, normally aspirated and super/turbocharged reciprocating IC engines, turbojets, turbofans, ramjets, pulsejet and pulse detonation engines, rockets, and even flapping and rotary wing systems. The details of most of these propulsion systems is beyond the scope of the typical introductory course in aerospace engineering, and further discussions will be limited to the details of a typical prop-driven IC engine propulsion system and only a brief glance at jet engines.

PROPELLER-DRIVEN PISTON-CYLINDER PROPULSION SYSTEMS

Nearly all light aircraft use propeller-driven propulsion systems equipped with IC engine powerplants. The basic system relies on the conversion of energy stored in the form of chemical bonds in hydrocarbon fuels into mechanical energy within the engine and drivetrain which is then converted through the propeller and into the surrounding air to take the form of kinetic energy. The introduction of new kinetic energy into the flow surrounding the aircraft then changes the momentum of the air and imparts a force in the form of forward thrust into the aircraft. Naturally, each energy conversion process occurs in a less-than-ideal manner, with efficiencies ascribed to each process to keep track of the loss of useful work moving from the fuel in the gas tank into forward thrust. The three most common efficiencies associated with such a propulsion system are the thermal, mechanical, and propeller efficiencies expressed as

$$\eta_{th} = \frac{\dot{W}_{out}}{\dot{Q}_{in}} \tag{6.4}$$

$$\eta_{mech} = \frac{\dot{W}_{out}}{\dot{W}_{in}} \tag{6.5}$$

$$\eta_{prop} = \frac{\Delta \dot{KE}_{air}}{\dot{W}_{in}} \tag{6.6}$$

The lowest of all these values is thermal efficiency, η_{th}, which is typically in the neighborhood of 20-35%. Thermal efficiency remains low for IC engines due to the limits in maximum temperatures and practical compression ratios achievable by such designs. Additional reductions in thermal efficiency are caused by the inability of the engine to take in a fresh charge of air at its ambient specific volume or expel exhaust as easily as possible leading to increased work load with reduced power output. Depending on the exact drivetrain, mechanical efficiency, η_{mech}, is often the highest value of the bunch. Mechanical efficiency accounts for the losses in power as the lubricated gears, shafts, and associated bearings move together, transferring the output rotary motion of the engine to the propeller at the hub. Typical values for mechanical efficiency range from 90-95%, again depending upon the specific drivetrain. Propeller efficiency accounts for the losses in directional kinetic energy as the propeller "screws" through the air (note propellers are still sometimes referred to as "airscrews"). Ideally, all the air motion induced by the movement of the propeller blades would propel the air in the axial direction relative to the propeller plane. In reality, some of the air motion is directed radially and tangentially to create large swirling vortices and prop wake disturbance. Typical values for propeller efficiency range from 75-90%.

The overall efficiency, η_o, of a propeller-driven propulsion system can be found as

$$\eta_o = \eta_{th}\eta_{mech}\eta_{prop} = \frac{\Delta \dot{KE}_{air}}{\dot{Q}_{in}} \tag{6.7}$$

Combining all the lowest and highest ranges for the individual efficiencies, one may find

$$13\% < \eta_o < 30\%$$

Examining the propeller flow in more detail, it is possible to show how the contracting tube of air moves through the propeller disk, increasing in speed as it moves. In the ideal case shown in Figure 6-6, air moving at the vehicle flight speed is brought through the propeller disk as a result of the lower pressure on the forward-facing surface of each propeller blade. At the propeller, the local velocity is increased by the induced velocity, w_i. In the far wake of the propeller, the propwash continues to contract and stabilizes once the wake velocity, V_w, is equivalent to the sum of the freestream velocity and twice the induced disk velocity.

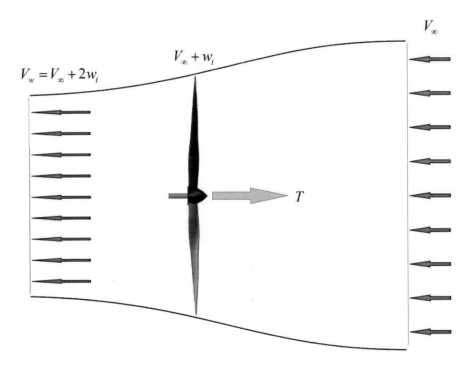

Figure 6-6: Upstream air contracts and speeds up as it moves through the spinning propeller disk.

Ideally the propeller wake velocity is uniform and the propeller shaft torque does not produce angular momentum in the wake. In that case the required ideal shaft power is found as

$$P_{shaft,ideal} = \Delta KE_{air} = \tfrac{1}{2}\dot{m}\left(V_w^2 - V_\infty^2\right) = \frac{d}{dt}\left(W_{prop}\right)$$

Consider that the ideal shaft output power is perfectly transferred to the propeller and into the air. This power can also be considered in its fundamental form as a rate of work, where the propeller work done is simply

$$W_{prop} = \int T dx$$

In terms of a rate of work, it is possible to take the time rate of change of the propeller work to find the output power as

$$P_{shaft,ideal} = TV_\infty \tag{6.8}$$

In reality the wake is non-uniform and has unintentional tangential and radial kinetic energy, as shown in Figure 6-7. This flow field results in a reduced propeller efficiency and resultant drop in transferred power such that

$$\eta_{prop} = \frac{TV_\infty}{P_{shaft}} < 1.00 \tag{6.9}$$

Note that the combination of Eq. (6.8) and Eq. (6.9) leads to the conclusion that the ideal shaft power is always greater than the actual output shaft power.

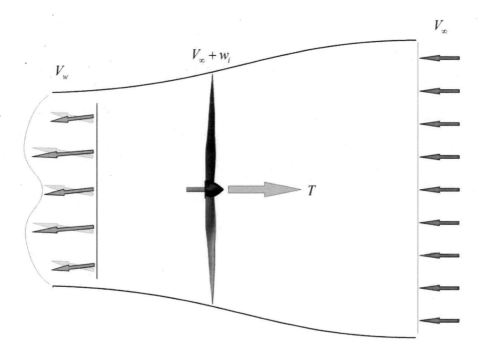

Figure 6-7: Actual propeller flow contains axial, tangential, and radial flow components, effectively reducing propeller efficiency.

Historically, propellers and airscrews have seen a fair amount of both good and bad attempts to produce more efficient mechanical-to-kinetic energy transfer. The key to a finely-designed propeller is the proper application of the airfoil such that the section lift at the specific radial location on the blade is tuned to the incoming air. It is important to understand that the relative velocity, V_R, varies depending upon the radial location of the blade. The greater the distance from the hub's center, the faster the speed and the shallower the relative angle of attack of the local wing section. This fact necessitates placing twist and taper on the propeller blade to maintain efficient attached flow everywhere along the radial span of the blades. Typically, the angle of incidence is quite high at the hub and nearly zero at the tip. Figure 6-8 shows how the rotational speed, ω, impacts the magnitude and direction of the relative velocity at a selected radial section of a typical propeller blade. The wing section produces lift and drag as it normally would, the results of which contribute to the forward thrust and torque exerted on the propeller's bub. The axially-aligned components of the lift vector and drag vectors produce thrust while the tangentially-aligned components of the lift and drag vectors multiplied by the local radius produce the torque on the propeller shaft, which the engine must provide to keep the propeller turning. The propeller produces wing tip vortices, just like a 3-D wing, except that they spiral around the wake.

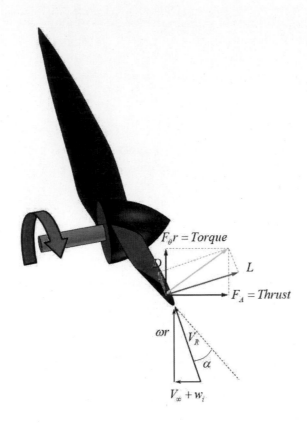

Figure 6-8: A schematic view of a radial section of a propeller blade shows the effect of spin on the relative velocity, V_R.

Referring to Figure 6-8, one may calculate the thrust as

$$T = F_A = L_A - D_A$$

and the torque as

$$Torque = F_\theta r = \left(L_\theta + D_\theta\right)r$$

The required power at the shaft is the integrated product of torque and rotational speed such that

$$P_{shaft} = \int_0^{r_{tip}} F_\theta\left(r\right)\omega\,dr$$

Furthermore, assuming negligible radial velocity, the relative velocity of the propeller section is found as the vector sum of the axial and tangential velocities, or

$$\vec{V}_R = \left(\vec{V}_\infty + \vec{w}_i\right) + \bar{\omega}r \tag{6.10}$$

The magnitude of the relative velocity is thus

$$V_R = \sqrt{\left(V_\infty + w_i\right)^2 + \left(\omega r\right)^2} \tag{6.11}$$

GAS-TURBINE ENGINE PROPULSION SYSTEMS

Propellers are shaft driven by piston engines up to 500 hp or by turboprops up to 2000 hp. The thrust decreases with ambient air density which limits their operating ceiling. The piston-cylinder engine available power, P_A, decreases linearly with ambient air density and differs from a jet engine in that its power is independent of flight velocity, thus the thrust is maximum on take-off and decreases with increasing flight speed. There is a simple way to improve the operational altitude of a propeller-driven propulsion system: high power level propellers are turbine driven and are called turboprops. These are jet engines with an extra turbine installed to drive the propeller via a reduction gear. While turboprops have ample power, this also decreases linearly with density at altitude, but because the turboprop has more power at sea level, it operates at higher ceilings than a piston engine driven propeller.

Above Mach 0.4 the efficiency of the conventional propeller decreases due to tip losses. It is not uncommon for propeller tips to exceed local Mach numbers of 1.2. However, when the propeller is surrounded by a shroud its blades behave like 2-D airfoils and the induced drag is eliminated. In this arrangement, the propeller is called a fan, which is efficient up to Mach 0.8. For higher flight Mach numbers, a combination of a fan and simple jet engine is most efficient. Unlike the turboprop a simple jet produces a nearly constant maximum jet thrust over a wide range of flight velocities (recall past homework problems such as problem 4.3) – how much will a 50% reduction in intake velocity impact the thrust? What about at higher speeds?) As thrust equals the increase in momentum, the decrease in velocity difference is offset by the increase in mass flow rate with increasing flight speed.

Considering the ability for the different engine types to take in air, the air mass flow rate through a jet engine is up to 50 times higher than through a piston engine with the same frontal area, explaining their high thrust-to-weight ratio. For example the General Electric F404-GE-FID turbojet has thrust 10,000 lb$_f$ and weighs 1730 lb$_f$ for a thrust-to-weight ratio of 5.8. In terms of thrust specific fuel consumption, s.f.c., a jet engine s.f.c. is about 0.6 lb$_f$ of fuel per hour per lb$_f$ thrust generated, as compared to 0.5 for a piston engine and propeller combination. A basic jet engine is actually a hot high pressure gas generator. The gas generator works by first compressing the air in the compressor stage(s), then heating the air by burning about 2% by mass of aviation jet fuel. About half of the pressure of this hot gas is lost during expansion in the turbine(s) which drive(s) the compressor(s). The enthalpy of the heated gas is finally converted to kinetic energy in the nozzle. Figure 6-9 shows a simple schematic of a jet engine.

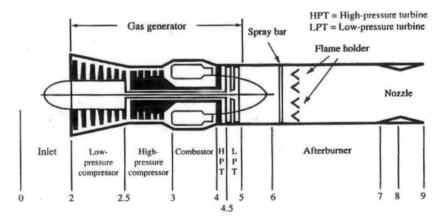

Figure 6-9: A jet engine uses compressors, combustors, turbines, and nozzles to create high velocity gas.

APPLYING POWER TO FLIGHT

For any case other than an unpowered glider, aircraft must have propulsion systems onboard in order to maintain force equilibrium during the many phases of flight. Consider the three general pitch attitude cases depicted in Figure 6-10:

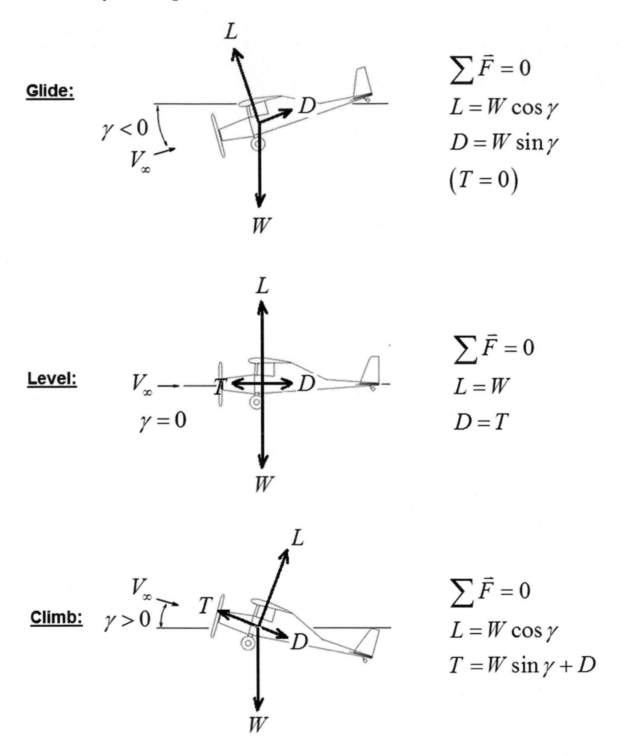

Figure 6-10: The three general cases of aircraft pitch attitude are shown with their respective forces.

When considering the forces about an aircraft, it is often helpful to use the defined aircraft axes rather than considering forces with respect to the Earth, thus "downward" indicates a direction aligned with the vertical axis, and is not necessarily aligned with the direction of gravity. With the aircraft reference axes in mind, in the case if gliding descent, drag is equalized by the forward component of weight and the downward component of weight is equalized by lift. In level flight, drag is equalized by thrust and weight is equalized by lift. In the event of a climb, the downward component of weight is equalized by lift, and the sum of the backward component of weight and the drag is equalized by the thrust.

No matter what the case, work (the product of force and distance) must be done as the aircraft moves through the air. Over time, this translates into an amount of power (work per unit time) required to move the vehicle. There are several specific powers that relate to prop-driven aircraft flight:

thrust power	$\Rightarrow$	TV_∞	(power to actually move the aircraft)
shaft power	$\Rightarrow$	P_{shaft}	(power delivered to the engine shaft)
required power	$\Rightarrow$	P_{req}	(power needed at the shaft for given D)
available power	$\Rightarrow$	P_A	(total shaft power being output by engine)
excess power	$\Rightarrow$	$P_{excess} = P_A - P_{req}$	(power being used for add'l climb needs)
maximum power	$\Rightarrow$	P_{max}	(total possible engine output at the shaft)

An additional parameter which is important for any non-level flight is the vertical flight speed with respect to the direction of gravity which is also commonly referred to as the aircraft rate of climb, R/C

vertical speed	$\Rightarrow$	V_{vert}	$\left(\begin{array}{l}\text{aircraft vertical speed with respect to } \bar{g}, \\ \text{aka "R/C" or "climb rate"}\end{array}\right)$

Now consider how each of these parameters may be used in the three aircraft configurations:

Glide: During a power-off glide, the thrust required to overcome the drag is readily available in the form of the forward component of the vehicle weight, so

$$"T" = W \sin\gamma = D$$
$$L = W \cos\gamma$$
$$P_{req} = 0$$

Furthermore, the glide angle and vertical speed may be found as

$$\tan\gamma = \frac{W \sin\gamma}{W \cos\gamma} = \frac{D}{L} \quad \Rightarrow \quad \gamma = \tan^{-1}\left(\frac{D}{L}\right)$$
$$V_{vert} = V_\infty \sin\gamma$$

Level: Here, the aircraft pitch is neutral and thrust will be needed, so

$$T = D$$
$$L = W$$
$$P_{req} = P_{shaft} = \frac{TV_{\infty}}{\eta_{prop}}$$

and the glide ratio and vertical speed are of course

$$\gamma = 0°$$
$$V_{vert} = 0 \left(m/\sec, etc. \right)$$

<u>Climb:</u> Here, the aircraft is pitched up and will most definitely need some thrust to overcome drag and also to provide work to lift the weight of the aircraft up, so

$$T = W \sin \gamma + D$$
$$L = W \cos \gamma$$

$$P_A = P_{req} + P_{excess} \qquad \left(P_{req} = \frac{DV_{\infty}}{\eta_{prop}}; \qquad P_{excess} = \frac{W \sin \gamma V_{\infty}}{\eta_{prop}} = \frac{WV_{vert}}{\eta_{prop}} \right)$$

The climb angle and vertical speed are found as

$$\gamma = \sin^{-1} \left(\frac{V_{vert}}{V_{\infty}} \right)$$
$$V_{vert} = V_{\infty} \sin \gamma$$

Generally speaking, fossil fuel oxidation (combustion) is the primary method of aviation power generation due to its extremely high energy density. The net work production from an internal combustion engine, however, is limited by its chemical reaction process. In particular, the density of the air arriving to the combustion chamber (usually a piston-cylinder assembly) is directly related to the amount of oxygen being delivered for each "power stroke" burn. For this reason, the maximum power deliverable by a normally aspirated engine varies linearly with ambient air density, or

$$P_{max} \propto \rho_{amb}$$

It will be shown later that this fact leads inevitably to what is known as the aircraft service ceiling which is an altitude above which the aircraft cannot maintain level flight.

6.5   BANKED TURNS

One of the most common ways to change heading in an aircraft is to make what is known as a banked turn. To perform this maneuver, the pilot begins by rolling off-level and then pulling up on the yoke to "climb" about a gentle horizontal arc. Consider a rear view of the aircraft during a banked turn as shown in Figure 6-11:

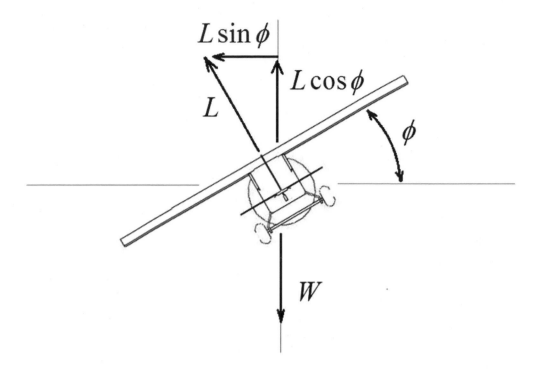

Figure 6-11: A light aircraft in a banked turn rolls off-level before "climbing" into the turn.

To maintain altitude, all the forces aligned with the direction of gravity must be in equilibrium, so

$$L \cos \phi = W$$

The careful student will observe, however, that there remains an unbalanced force, namely

$$L \sin \phi \neq 0$$

which causes the vehicle mass to accelerate laterally. This can be better understood by examining the situation from a view of the top of the aircraft as shown in Figure 6-12:

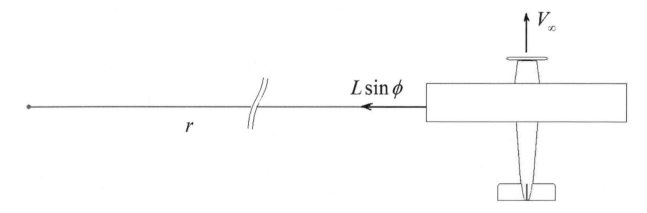

Figure 6-12: A top-view of an aircraft in a banked turn shows the critical parameters related to the physics of the maneuver.

179

From the parameters shown in Figure 6-12, it is now possible to solve for the banked turn radius, using the mass and centripetal acceleration of the vehicle as key characteristics so that

$$a_c = \frac{V_\infty^2}{r} = \frac{gL\sin\phi}{W}$$

Solving for the turn radius, r,

$$r = \frac{V_\infty^2 W}{gL\sin\phi} \tag{6.12}$$

It is helpful to now define a load factor (sometimes called the "g-factor"), n, such that

$$n = \frac{L}{W} = \frac{1}{\cos\phi} \tag{6.13}$$

It is now possible to rewrite Eq. (6.12) to express the turn radius in terms of the load factor,

$$r = \frac{V_\infty^2}{g\sqrt{(n^2-1)}} \tag{6.14}$$

6.6   ENDURANCE AND RANGE

Of all of the performance parameters an aircraft has associated with it, perhaps the most fundamentally important aspects in overall mission application are the flight vehicle range and endurance. Range, R, is usually provided in feet or meters and indicates how far the aircraft may travel in the air before needing to land. Endurance, E, is usually provided in hours and indicates how long the aircraft may remain aloft before needing to land as a result of dangerously low fuel. In either case, an aircraft is usually designed to maintain a fuel capacity of 10% reserves to make sure that unexpected emergencies or extra loiter times may be handled.

One might naturally assume that range and endurance go hand-in-hand and that to maximize one should lead to the maximization of the other. Curiously enough, however, the conditions required to maximize range are not the same as those required to maximize endurance. Furthermore, aircraft missions may make one of the two design parameters more important than the other. For example, Intelligence, Surveillance, and Reconnaissance (ISR) missions will almost certainly opt for enhanced endurance. Commercial air transport, on the other hand, will almost always seek to maximize range. This section will briefly discuss how to calculate both endurance and range for a given aircraft. Note that these calculations are only valid for aircraft with prop-driven piston-cylinder engine propulsion systems.

ENDURANCE

One of the most important things for a pilot to know is how long the aircraft is able to remain in the air. This is called the aircraft endurance, E, which is measured in hours.

$$E(hr) = C_E \left(\frac{C_L^{1.5}}{C_D} \right)$$

(6.15)

Note that the coefficient of endurance, C_E, has units of hours and is calculated as

$$C_E = \frac{\eta_{prop}}{3600c} \sqrt{(2\rho_\infty S_w)} \left[\frac{1}{\sqrt{(W_0 - W_f)}} - \frac{1}{\sqrt{W_0}} \right]$$

(6.16)

Where c is the fuel consumption rate for propeller-driven aircraft,

$$c = 2.42 \times 10^{-7} \left(per\ ft \right) = 8 \times 10^{-7} \left(per\ m \right)$$

and W_0 is the initial aircraft takeoff weight and W_f is the weight of the fuel used during flight.

To maximize endurance, one may simply evaluate the expression with respect to the lift coefficient. It is possible to show that the global maximum occurs for the endurance expression when

$$C_L = \sqrt{3C_{D,para}\pi eAR} \quad \Rightarrow \quad E = E_{max}$$

RANGE

Like endurance, pilots are also interested in knowing how far the aircraft may travel before having to land. The expression for range, R, is

$$R = C_R \left(\frac{C_L}{C_D} \right)$$

(6.17)

Where

$$C_R \left(ft\ or\ m \right) = \frac{\eta_{prop}}{c} \ln \left(\frac{W_0}{W_0 - W_f} \right)$$

(6.18)

Note that the coefficient of range, C_R, has units of feet or meters depending upon the selection of fuel consumption rate, c.

To maximize range, one needs only to maximize the ratio of lift to drag coefficients, but this condition is already known:

$$C_L = \sqrt{C_{D,para}\pi eAR} \quad \Rightarrow \quad R = R_{max}$$

6.7 TAKE-OFF AND LANDING DISTANCES

One of the limiting factors for any aircraft design is that it must be able to take-off and land within the lengths of existing runways. Aircraft weight, wing loading, maximum takeoff thrust, and maximum lift coefficient are of primary concern in ensuring short enough lengths for appropriate airport runways. For example, a fully-loaded Airbus A 380 with 1.24 MN thrust weighing nearly 5.64 MN loaded on a wing with an area of 845 m² (for a wing loading of 1470 Pa) requires 2950 m to lift off; a Piper J3 Cub with a 2.5 kN thrust weighing 5.4 kN loaded on a wing of just under 16.6 m² (for a wing loading of 325 Pa) requires 222 m to lift off.

TAKE-OFF

One might first think that the minimum velocity needed to take off is simply the calculated stall speed of the aircraft. This is not in fact the case, since the airplane must develop enough lift to overcome weight and climb. For this reason, the accepted value of takeoff speed is 20% higher than stall speed, or

$$V_{TO} = 1.2V_{stall} \qquad (6.19)$$

Flaps may be deployed to increase $C_{L,max}$ which reduces the stall velocity and thus reduces takeoff speed. This effectively reduces the distance required to takeoff since the aircraft can reach the lower speed sooner. Note that once the vehicle is safely away from the ground and all potential obstructions, the flaps must be taken in to reduce the substantial drag penalty imposed by the large structures on the wings. To estimate the takeoff distance, consider an aircraft on the runway preparing for takeoff and identify all the force components acting on the vehicle, as shown in Figure 6-13.

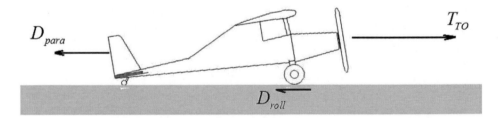

Figure 6-13: An airplane ready to takeoff is shown with all the force vectors acting on it.

As the aircraft begins to speed up, the parasitic drag increases, but the rolling drag and the thrust remain constant, so the net force and thus the net acceleration on the aircraft are not constant during takeoff. To simplify the problem, assume a constant average acceleration, such that

$$a_{TO,avg} = \frac{F_{TO,avg}}{W_0}g \qquad (6.20)$$

To find the average accelerating takeoff force, $F_{TO,avg}$, one must assume an average velocity of the entire takeoff event. It is possible to show that the constant velocity required to achieve an equivalent average takeoff force during a takeoff event which actually has variable force is approximately 70% of the maximum velocity, or

$$V_{TO,avg} = 0.7V_{TO} \tag{6.21}$$

Once the average takeoff velocity is known, the average dynamic pressure during takeoff can be calculated as

$$q_{TO,avg} = \tfrac{1}{2}\rho_\infty \left(0.7V_{TO}\right)^2$$

The accelerating force balance during takeoff is

$$F_{TO} = T_{TO} - D_{para,TO} - D_{roll,TO} \tag{6.22}$$

During an actual takeoff event, the takeoff thrust will vary as forward speed increases, but the average takeoff thrust may be determined from the average thrust power during takeoff such that

$$T_{TO} = \frac{\eta_{prop}P_{max}}{\left(0.7V_{TO}\right)} \tag{6.23}$$

The power used during takeoff is usually at or near maximum power as shown in Eq. (6.23) for obvious reasons.

With the average takeoff dynamic pressure known, it is relatively easy to find the average parasitic drag as

$$D_{para,TO} = C_{D,para}q_{TO,avg}S_w \tag{6.24}$$

The rolling drag is a result of the friction between the runway and tires, the wheel bearings, and even the brakes when the pilot applies them during landing. The values of the rolling drag depend a great deal on whether the brakes are applied, of course, but in general,

$$D_{roll} = \mu_{roll}W \tag{6.25}$$

During takeoff, the brakes are not applied and the aircraft is at its heaviest weight; during landing, the brakes are applied and the aircraft is at its lightest occupied weight, so

$$\textit{takeoff} \quad \Rightarrow \quad W = W_0; \qquad \mu_{roll} \in [0.02, 0.05] \,(\textit{free roll})$$

$$\textit{landing} \quad \Rightarrow \quad W = (W_0 - W_f); \qquad \mu_{roll} \in [0.2, 0.5] \,(\textit{brakes applied})$$

Now that the average takeoff acceleration may be calculated, it is possible to integrate

$$a = \frac{dV}{dt} \quad \Rightarrow \quad \int_0^{TO} a\,dt = \int_0^{TO} dV \quad \Rightarrow \quad a_{TO,avg}t_{TO} = V_{TO}$$

$$V = \frac{ds}{dt} \quad \Rightarrow \quad \int_0^{TO} V\,dt = \int_0^{TO} ds = \int_0^{TO} a_{TO,avg}t\,dt$$

to find the takeoff distance, s_{TO},

$$s_{TO} = \tfrac{1}{2} a_{TO,avg} t^2 = \frac{1}{2}\left(\frac{V_{TO}^2}{a_{TO,avg}} \right) \qquad (6.26)$$

Combining Eq. (6.19), Eq. (6.20), and Eq. (6.26), it is possible to estimate the takeoff distance as

$$s_{TO} = \frac{0.72 V_{stall}^2 W_0}{F_{TO,avg}\, g} \qquad (6.27)$$

A head wind, V_{hw}, which is defined as wind coming into the nose of the aircraft, can help shorten the required takeoff distance because it adds "free" dynamic pressure. In the presence of a headwind, the new takeoff distance, $s_{TO,hw}$, may be calculated using a simple modification to the wind-free takeoff distance of Eq. (6.27) such that

$$s_{TO,hw} = s_{TO}\left[\frac{\left(V_{TO} - V_{hw}\right)}{V_{TO}} \right]^2$$

In the event of a tail wind, simply use a negative value for the headwind to find the new takeoff distance. It should be clear to even the casual observer how important the headwind velocity can be in safe takeoff or landing.

LANDING

When landing, similar arguments must be made about the slowing acceleration ("deceleration" in the vernacular) being some kind of an average. Again, one needs to start with a basic maximum velocity. The accepted value for safe landing speed is 30% higher than stall speed, or

$$V_L = 1.3 V_{stall} \qquad (6.28)$$

The average aircraft during a landing event is

$$V_{L,avg} = 0.7 V_L \qquad (6.29)$$

The associated average landing dynamic pressure is

$$q_{L,avg} = \tfrac{1}{2}\rho_\infty \left(0.7 V_L\right)^2$$

Now considering the forces acting on the aircraft while it is landing, noting that this time there is no need for thrust, one may show all the forces resolved about the aircraft as shown in Figure 6-14

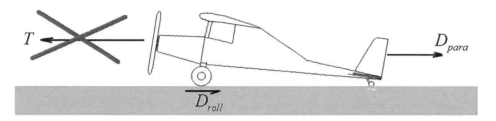

Figure 6-14: An airplane in the process of landing is shown with its force vectors. Note there is no need for thrust during landing since forward acceleration is undesirable.

The force balance describing the aircraft during landing is

$$F_L = D_{para,L} + D_{roll,L}$$ (6.30)

The parasite drag is found the same way as usual,

$$D_{para,L} = C_{D,para} q_{L,avg} S_w$$ (6.31)

The rolling drag is found using Eq. (6.25), with the higher value of brake friction coefficient being used (since the brakes are being applied) and the weight of the aircraft reduced to take into account the fact that a significant weight in fuel has been expended during normal flight prior to landing.

For the "decelerating" landing process,

$$a_{L,avg} = \frac{F_{L,avg}}{(W_0 - W_f)} g$$ (6.32)

And finally, the landing distance may be estimated as

$$s_L = \frac{0.845 V_{stall}^2 (W_0 - W_f)}{F_{L,avg} g}$$ (6.33)

6.8   PERFORMANCE AT ALTITUDE AND CEILING

Air breathing engines are always limited in their power output by the combustion stoichiometry, thus the ambient air density plays a crucial role in determining the total power output possible for a given aircraft power plant. By assuming a linear reduction in air density (thus power output) with altitude, one can simply determine what is known as the aircraft service ceiling. The service ceiling is defined as the altitude at which an aircraft can climb at a maximum rate of 100 feet per minute. The absolute ceiling is defined as the altitude at which an aircraft can no longer climb at all. An example problem is perhaps the best way to show how service ceiling is determined for a given aircraft.

EXAMPLE 6.1: FINDING THE ABSOLUTE AND SERVICE CEILINGS FOR AN AIRCRAFT

Given:

Consider an aircraft which has a known sea level climb rate and a known climb rate at a given altitude during flight.

$$\left(R/C\right)_{sl} = 1500 \; ft/min \qquad\qquad \left(R/C\right)_{9k} = 400 \; ft/min$$

Find:

a. The aircraft service ceiling
b. The aircraft absolute ceiling

Solution:

It will help to start with a sketch of the problem. Next use similar triangles to solve:

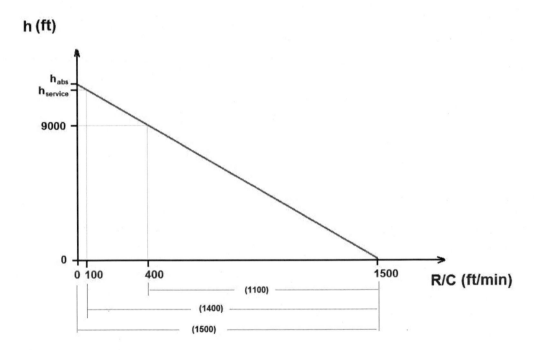

a)

$$\frac{1100}{9000} = \frac{1400}{h_{service}} \qquad \Rightarrow \qquad h_{service} = 9000 \; ft\left(\frac{1400}{1100}\right) = 11,454 \; ft$$

b)

$$\frac{1100}{9000} = \frac{1500}{h_{abs}} \qquad \Rightarrow \qquad h_{abs} = 9000 \; ft\left(\frac{1500}{1100}\right) = 12,272 \; ft$$

Check:

The answers seem reasonable in magnitude and have the right units.

6.9   SUMMARY

AIRCRAFT ATTITUDE

All aircraft follow a basic plane-fixed coordinate system. Table 6-2 Summarizes:

Table 6-2: The basic aircraft coordinate system.

Axis	Associated Translational Freedom and Sign Convention	Associated Rotational Freedom and Sign Convention
x (longitudinal, roll)	(+) Forward (-) Backward	(+) Roll CW – starboard wingtip downward (-) Roll CCW – port wingtip downward
y (lateral, pitch)	(+) Sideslip starboard (-) Sideslip port	(+) Pitch up - nose up (-) Pitch down – nose down
z (vertical, yaw)	(+) Descend (-) Ascend	(+) Yaw CW – port wingtip forward (-) Yaw CCW – starboard wingtip forward

POWERED FLIGHT

For a prop-driven aircraft, the power from the engine is reduced as it makes its way to the passing air. The propeller is not absolutely perfect in efficiency, so

$$\eta_{prop} = \frac{\Delta KE_{air}}{P_{shaft}} = \frac{TV_\infty}{P_{shaft}} < 1.00$$

There are three conceivable pitch attitudes that impact power: power-off glide, level, and climb. The relevant cats for each condition are shown in Table 6-3:

Table 6-3: Pitch attitudes and resulting equilibrium conditions.

	Power-off Glide	Level Flight	Climb
L	$W \cos \gamma$	W	$W \cos \gamma$
D	$W \sin \gamma$	T	$T - W \sin \gamma$
γ	$\tan^{-1}\left(\dfrac{D}{L}\right)$	0	$\sin^{-1}\left(\dfrac{V_{vert}}{V_\infty}\right)$
V_{vert}	$V_\infty \sin \gamma$	0	$V_\infty \sin \gamma$
P_{req}	0	$P_{shaft} = \dfrac{TV_\infty}{\eta_{prop}}$	$\dfrac{DV_\infty}{\eta_{prop}}$
P_{excess}	0	0	$\dfrac{WV_{vert}}{\eta_{prop}}$

BANKED TURNS

For a banked turn where the aircraft is at a bank angle, ϕ, the turn radius, r, is

$$r = \frac{V_\infty^2}{g\sqrt{(n^2-1)}}; \qquad n = \frac{1}{\cos\phi}$$

ENDURANCE AND RANGE

Range is how far an aircraft can fly before having to land. Endurance is how long an aircraft can fly before having to land. All relevant equations are summarized in Table 6-4.

Table 6-4: Endurance and range summary.

	Endurance	Range
Basic form	$E(hr) = C_E\left(\dfrac{C_L^{1.5}}{C_D}\right)$	$R = C_R\left(\dfrac{C_L}{C_D}\right)$
Coefficient	$C_E = \dfrac{\eta_{prop}}{3600c}\sqrt{(2\rho_\infty S_w)}\left[\dfrac{1}{\sqrt{(W_0-W_f)}} - \dfrac{1}{\sqrt{W_0}}\right]$	$C_R(ft\ or\ m) = \dfrac{\eta_{prop}}{c}\ln\left(\dfrac{W_0}{W_0-W_f}\right)$
Max condition	$C_L = \sqrt{3C_{D,para}\pi eAR}$	$C_L = \sqrt{C_{D,para}\pi eAR}$

$Note: \quad c = 2.42\times10^{-7}\ (per\ ft) = 8\times10^{-7}\ (per\ m)$

TAKEOFF & LANDING, AND SERVICE & ABSOLUTE CEILING

Takeoff and landing characteristics are critical in terms of being able to use existing runways. A summary of equations for both events are summarized in Table 6-5.

Table 6-5: Takeoff and landing summary.

	Takeoff	Landing
V	$1.2V_{stall}$	$1.3V_{stall}$
F	$T_{TO} - D_{para,TO} - D_{roll,TO}$	$D_{para,L} + D_{roll,L}$
T	$\dfrac{\eta_{prop}P_{max}}{(0.7V_{TO})}$	0
D_{roll}	$\mu_{roll}W_0; \qquad \mu \in [0.02, 0.05]$	$\mu_{roll}(W_0-W_f); \qquad \mu \in [0.2, 0.5]$
s	$\dfrac{0.72V_{stall}^2 W_0}{F_{TO,avg}\,g}$	$\dfrac{0.845V_{stall}^2(W_0-W_f)}{F_{L,avg}\,g}$

Service ceiling occurs where R/C is 100 feet per minute. Absolute ceiling occurs where R/C is 0.

CHAPTER 6 PROBLEMS

6.1) Using the information presented in Table 6-1, determine the validity of Eq. (6.3) by comparing the actual to the estimated wing loading as it applies to the weight of the following flyers:

    a. Dragonfly
    b. Turkey Vulture
    c. Boeing 747-400

6.2) Explain why a propeller blade must have twist and taper to make it efficient.

6.3) Use at least one conservation principle discussed in Chapter 4 to explain the primary reason for a propeller blade's apparent loss of energy transfer when generating thrust.

6.4) A Piper J-3 Cub can deliver 48 hp to the propeller shaft at an altitude of 10,000 ft. The light aircraft is equipped with a 79% efficient propeller that can propel it at a max speed of 87 mph at that altitude. The J-3 weighs 1,220 lb_f. Calculate the following:

    a. Find the drag of the aircraft when it is traveling at max speed at 10,000 ft.
    b. Find the value of the aircraft L/D when operating under that condition
    c. Assuming the engine power varies linearly with ambient density, estimate the power available to the propeller shaft at sea level (in hp).

6.5) Consider a single engine airplane with wing area of 15 m², a wing aspect ratio, $AR = 7$, Oswald efficiency, $e = 0.8$, and parasite drag coefficient, $C_{D,0} = 0.025$. Flying at 250 m/sec at a weight of 15,000 N, calculate the following:

    a. In level flight at sea level, find the lift, drag and thrust required (in N)
    b. At a 5 degree climb angle at sea level find the lift, drag and thrust required (in N)
    c. In level flight at 10,000 ft, find the lift, drag and thrust required (in N)
    d. At 5 degree climb angle at 10,000 ft, find the lift, drag and thrust required (in N)

6.6) Carefully show how one may obtain Eq. (6.14) using Eq. (6.12) and Eq. (6.13). Be sure to show each step for full credit.

6.7) An airplane is in a turn at $V_\infty = 175$ mph. Calculate the turn radius at various bank angles.

    a. 30 degree bank
    b. 60 degree bank

6.8) A Piper Seneca V must burn 732 lb_f of fuel to reach a max range of 920 miles. Its maximum takeoff weight is 4,750 lb_f, and its propellers are 87% efficient. Estimate the value of the aircraft's maximum L/D.

6.9) For the airplane in 6.5) at gross weight with a given $C_{Lmax} = 2.5$.

    a. Estimate the sea level takeoff distance and lift off speed on asphalt ($\mu = 0.02$)
    b. Estimate the sea level landing distance and touch down speed while braking ($\mu = 0.3$)

APPENDIX A: GLIDER DESIGN-BUILD-FLY PROJECT

GLIDER FOR CLASSROOM CONSTRUCTION

Constructing a simple glider is perhaps the best way to demonstrate a wide range of aspects of aerospace engineering while also showing the importance of each aircraft component. Test flying a simple glider helps students appreciate the need to balance forces and moments about the center of gravity of an airplane. An example of a simple foam-and-balsa glider is shown in Figure A-1.

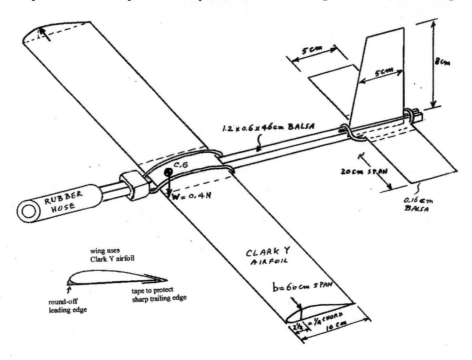

Figure A-1: A foam-and-balsa glider built in about one hour can help young aerospace engineers learn the value of proper weight distribution and wing geometry.

To fabricate such a glider, students work best in pairs. Two people are needed to handle a bow with an electrically-heated hot wire which is used to cut the wing out of blue construction foam. Sheet metal templates required to guide the hot wire around the upper surface of the wing are mounted on the ends of a 2 feet long wooden board. A wing geometry based on a Clark Y airfoil is preferred because it has a nearly flat bottom which is readily provided by the flat surface of the foam sheet. The wing is cut out upside down by two students holding the board down on a foam sheet while pulling the hot wire steadily through the foam along the template. To assure a straight trailing edge, students should begin cutting the foam from the trailing edge side. To ensure even entry of the hot wire into the foam surface, the metal template extends from the trailing edge to well above the foam surface. This allows a smooth sliding motion of the hot wire prior to entering the foam sheet.

Students should hand finish the wing surface prior to attaching the wing to a 0.5 x 0.25 x 18 inch long balsa or basswood fuselage with an elastic band. Drywall sanding paper or foam-on-foam sanding works well to smooth out the foam surface without gouging it. Students should be made aware of the need for a smoothly rounded leading edge and for a sharp trailing edge to optimize flight performance. Folding a strip of masking tape over the trailing edge helps to keep it sharp. Another strip of tape should be wrapped around the middle of the wing in the chordwise direction to minimize damage from the rubber band. The horizontal and vertical tail surfaces are cut from 2 inch wide 1/16th inch thick balsa wood using scissors or a hobby knife. The 0.5 inch wide fuselage needs a 2.5 inch long saw cut in the back, into which the vertical tail may be inserted. The horizontal tail is held in place by a short elastic band. A 2 inch long, 0.5 inch inside diameter heavy duty rubber hose is slipped over the nose of the fuselage to provide weight to balance the weight of the tail and to minimize the danger of hurting other students during test flights. Adding a ¼ inch long slice of this rubber hose is useful when placed against the wing leading edge to prevent it from sliding forward during each landing impact. Through trial and error, students learn that the glider only flies well if its center of gravity is near the wing's quarter chord position. This is because the center of wing-lift happens to occur near the quarter chord position. The glider shown in Figure A-1 can be expected to have a mass of about 40 grams and thus a weight of about 0.09 lbf.

It is critical to launch the glider with the appropriate velocity at a slight downward angle as shown in Figure A-2. If the glider is launched too fast, the nose will pitch up and the glider will enter a climb leading to a loss in air speed and an increase in angle of attack. This will inevitably cause the wing to stall and cause the glider to enter a potentially unrecoverable dive. The proper launch speed is found by trial and error till it flies furthest at a approximately constant glide speed, V_∞, and nearly constant glide slope, γ.

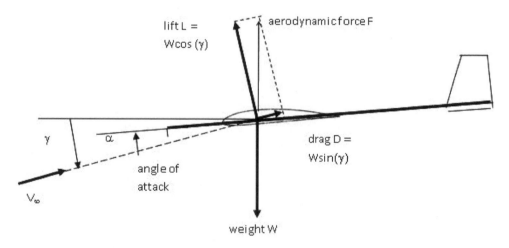

Figure A-2: A glider flies at a downward angle such that the forward component of its weight is equivalent to the aerodynamic drag it generates.

From flight tests, one will find the optimum velocity, $V_{\infty,opt}$, to be about 15 ft/sec and glide slope γ to be about 6°. As a glider does not have an engine to provide thrust, it must come down (except in the presence of thermals). In flight, the force component perpendicular to V_∞ is called lift, L, and the one in the direction of V_∞ is called drag, D. Their ratio is referred to as the lift-to-drag ratio, L/D. In a steady glide, all forces in the x and y directions must be balanced. The tangent of the glide angle must therefore equal the inverse of the lift-to-drag ratio. Mathematically, this relationship is written

$$\gamma = \tan^{-1}\left(\frac{D}{L}\right)$$

191

GLIDER DESIGN-BUILD-FLY TERM PROJECT

One of the best (and definitely one of the most fun) ways of demonstrating the fundamental knowledge students have learned in from this textbook is to design, build, and fly a hand-launched glider which can glide at least 40 feet when launched from a typical hand-held height (assume between 5 and 6 feet). Students are recommended to join in groups of up to three people to work throughout the semester in preparation for the Glider Fly-Off competition at the end of the term.

Three primary design constraints will drive the project:
1. The glider must weigh no more than 4 ounces.
2. The glider's wingspan must not exceed 3 feet.
3. For safety reasons, the nose of the glider must be constructed of a soft material.

Four glider-related assignments will be due during the semester and will count as 20% of the overall semester project grade:
1. Provide the names and contact info of each student on the project team.
2. Provide the first concept design of the glider, a list of materials to be used (with sources), and a project Gantt chart indicating the team's project schedule.
3. Participate in the Classroom Glider Construction Day.
4. Provide a dimensioned CAD three-view drawing, a bill of materials (BOM), and the anticipated glider weight along with its calculated center of gravity.

Students should feel free to communicate with the instructor as often as they like with questions regarding any aspect of the project (e.g., "Is it OK to spray paint my freshly-cut foam wing?", "Where should we put our glider's center of gravity?", "Are knives mounted on the nose OK?", etc.). Seeking additional information (e.g., textbooks, other faculty) is acceptable AS LONG AS the final design, analysis, construction, and other required work is solely the work of the team members.

The glider project demo and report will make up 80% of the term project grade, graded as follows:
- Presentation (20 pts): Based on quality of construction, originality of design, and application of fundamental ideas presented in lecture/text during the semester; at least three judges will review and/or discuss each design with team leaders before the glider is thrown; grade will be determined by averaging scores given by the judges.
- Performance (20 pts): Based on glider's ability to achieve a free-flight distance of at least 20 ft. (>40 ft. = 20 pts., >30 ft. = 15 pts., >20 ft. = 10 pts., <20 ft. = 0 pts.)
- Report (40 pts.): Based on completion/quality of required sections as follows:
  - Title page (include team #, team members)
  - Table of Contents
  - Design constraints and objective(s) (present in well-proofed paragraph form)
  - Cost analysis (can be tabulated, list source and cost of each material used)
  - Tabulated glider specifications (include specs such as aspect ratio, weight, wingspan, airfoil used, etc.)
  - Airfoil selection and analysis (must include technical discussion and include relevant graphics; use of XFLR-5 or other numerical analysis tool is mandatory; comparison of numerical results to existing experimental data is highly encouraged)
  - 3-view drawing of final design (*must use CAD*, drawn to scale and neatly and completely dimensioned)

     o   Tabulated and coded performance analysis with graphical presentation of results (*must use numerical analysis and spreadsheet tools*, e.g., MATLAB® and Microsoft® Excel®)

     o   Construction (present in well-proofed paragraph form)

     o   Flight testing results (present in well-proofed paragraph form)

     o   Images supporting the content of discussion sections are strongly encouraged

Glider Fly-Off will take place on the last Friday of class. Tables will be prepared along the walls of the launch area with team numbers on each table. Team leaders should arrive 15 minutes before class so that the completed glider and project report can be displayed at the appropriate table.

RECOMMENDATIONS FOR GLIDER DESIGN

1) Use balsa wood or foam materials for low weight and high strength. Especially vulnerable are the leading edge and the trailing edge of the wing and tail surfaces.
2) A smooth surface finish is important, especially for the main wing. If a shrink wrap is used, be careful not to overheat as it can shrink so much as to crush or deform the wing structure.
3) Be sure tail surface leading edges are not sharp but nicely rounded with $3/16^{th}$ radius or more
4) Make sure all trailing edges are sharp, since this aspect controls circulation by the Kutta condition and thus controls the maximum obtainable lift coefficient, C_L.
5) Although a high aspect ratio, AR, for the main wing promises higher L/D, this is only true if one is an expert builder. For beginners that have a higher likelihood of lower build quality, the increased scale of a low aspect ratio wing works in their favor. This means it can glide slowly like an owl. Note also that low AR wings are usually more resilient in crashes.
6) Estimate component weights (e.g., balsa wood has a specific gravity ranging from 0.2 to 0.4) and c.g. locations. This is easily done using a professional CAD program such as SolidWorks®.
7) In selecting the overall glider geometry, the following tips may be of use to beginning designers:
    a) Choose a fuselage length from 5 to 8 times the wing chord.
    b) Place the wing quarter chord from 1.5 to 2 times the wing chord length aft of the nose.
    c) Make the nose of the fuselage soft to both cushion the landing impact and mitigate damage to colleagues when they get hit. Warn all around prior to launch.
    d) Use NACA 0012 or similar symmetric airfoil for tail surfaces.
    e) To get correct tail volume use a horizontal tail with AR 4 to 5 located at 3 to 4 chord lengths aft of the c.g. with the corresponding horizontal tail area equal to 13% - 17% of the wing area. For the corresponding vertical tail area use 6% - 8% of the wing area.
    f) The corresponding horizontal tail volume V_h is calculated with $V_h = (S_h/S_w)(l_t/c)$ here ranging from 0.13 x 3 = 0.39 to 0.17 x 4 = 0.68.
    g) The corresponding vertical tail volume V_v is calculated as $V_v = (S_v/S_w)(l_t/c)$ here ranging from 0.06 x 3 = 0.18 to 0.08 x 4 = 0.32.
    h) Attach the wing and tail surfaces with elastic bands so they can break loose on impact without damage. Make sure that the wing attachment area is strong.
    i) To minimize risk of turning in flight and consequently hitting the walls before reaching the required distance, it is best to mount the wing on top of the fuselage to keep the glider c.g. below the wing. Teams can also give the wing dihedral of up to 6° to help maintain a straight flight. Make sure the wing is symmetric in shape, twist, and weight. An adjustable rudder is also recommended.

APPENDIX B: ADDITIONAL REVIEW PROBLEMS

CHAPTER 1-3 REVIEW PROBLEMS

1) When an aircraft is flying in steady level flight, it is said to be in equilibrium. Explain what this means mathematically and in words.

2) How does one "move" or "exchange" units between m-L-t and F-L-t approaches?

3) Helium is compressed inside a pressure vessel and held until it reaches thermal equilibrium with the room air. If the helium is allowed to suddenly rush into the room so that its bulk velocity is extremely high, what happens to its static temperature as it moves? Explain why this happens for full credit.

4) Why do pilots use temperature or pressure altitudes when describing their flight altitude?

5) Convert 345 °F into °C.

6) Convert 125,000 Pa into mmHg.

7) On planet Loth, located in a galaxy far, far away, the radius of the planet is r_p = 4500 km and the sea level gravitational acceleration is g_0 = 7.21 m/sec². The Lothian atmosphere is made up of a toxic gas with a known specific gas constant, R_{gas} = 352 N-m/kg-K. If the entire atmosphere is isothermal (temperature is constant throughout), with $T_{surface}$ = -19 °C and $p_{surface}$ = 154,000 Pa, find the following:

   a. What is the lapse rate of the atmosphere?
   b. What is the acceleration due to gravity at a geometric distance of 30 km from the surface?
   c. What is the density of the atmosphere at a geopotential altitude of 2 km?

8) A 55 gallon barrel is placed into water and loaded with weights until it just dips below the surface of the water. Given that 1 gal = 0.1337 ft³, the density of the air is 0.002377 slug/ft³ and the density of water is 1.94 slug/ft³, how much weight has been loaded onto the barrel?

9) A space station operates in low Earth orbit (LEO) at a geometric altitude of 300 km above the Earth's surface. Knowing that r_e = 6357 km and g_0 = 9.81 m/sec², find the following:

   a. What orbital velocity is required to maintain a steady orbit at that altitude?
   b. What is its period? (i.e., how much time is required for the space station to travel once around the Earth?)

10) A manometer is filled with a liquid which has a density six times greater than the density of water. If one end of the manometer is connected to a pressure vessel held at 22 psia and the other end is open to an ambient pressure of 12.3 psia, what will the difference in height in the two legs of the tube be in inches?

CHAPTER 4 REVIEW PROBLEMS

1) In a steady gas flow field, what is the maximum permissible time-rate-of-change of density within the flow? Show any necessary calculations or mathematical expressions.

2) At a given moment in time, an engine intake "breathes" in air at a rate of 0.0045 slugs/sec. If the downward motion of the piston has a speed of 30 ft/sec and the density of the air within the cylinder is 90% of standard sea level air, what is the bore size of the engine cylinder?

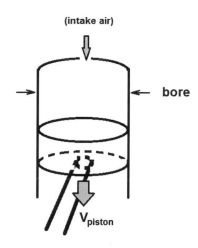

3) Your two-year-old loves to see if she can overflow the family's bathroom sink. With water entering the bathroom sink from the faucet at a maximum rate of 0.2 kg/sec, it is possible to get the water to rise to the safety overflow at the upper rim of the sink so that the water level just covers the hole and remains at that level. If the flow rate of the basin drain at the base of the sink is 3.2 times the flow rate of the overflow drain, what is the flow rate of water through the overflow?

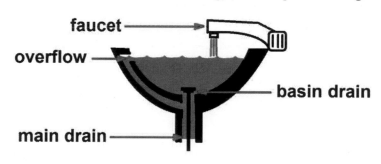

4) Your college professor told you that he used to build jet-powered devices to use in his home swimming pool by connecting a hose (with a 1 inch inner diameter) to one of the pool jets and then attaching the other end of the hose to different types of floating and submersible vehicles. If the pool pump delivers water to each pool jet at a rate of 100 gallons per minute, what is the maximum force the jet can deliver (Note $\rho_{H2O} = 1.94$ slug/ft³)?

5) You attach a compressed nitrogen bottle to your skateboard so that it is aligned with the free-rolling direction, strap your feet onto the contraption, and smash the end cap of the bottle off with a big hammer. If the mass flow rate of the exhaust gas is 73 kg/sec and the gas has a velocity of 350 m/sec, what is the thrust force generated **in lb_f** (assume the pressure thrust is negligible in your calculations)?

6) What is the speed of sound in desert air that has a static temperature of 104 °F?

7) What is the highest flow Mach number allowable before one must consider the compressibility of the flowing gas?

8) A Pitot probe attached to an RC airplane reads a flow total pressure of 1.043 atm while moving along through a Standard Atmosphere at sea level. What is the speed of the RC airplane? Use the Bernoulli equation for your solution.

9) What is the total temperature of air which has a velocity of 230 m/sec and a static temperature of 300 K?

10) During reentry, a spacecraft records a temperature of 1389.18 K on the nose of the craft at an exact geometric altitude of 30 km in a Standard Atmosphere. What is the Mach number of the spacecraft at that moment? Assume the nose of the spacecraft is the stagnation location in your calculations.

11) What are the two conditions required for a flow to be considered isentropic?

12) A commercial airliner flies at the cruise conditions indicated below. Find the values of the stagnation temperature and pressure within the flow given the following airspeed and static state variables:

$$V_{cruise} = V_\infty = 822 \; ft/sec; \qquad T_\infty = -70°F; \qquad p_\infty = 475 \; psfa$$

13) Recalculate the stagnation pressure of problem 12) using the Bernoulli equation if the air density is known to be 0.0007080 slug/ft³.

14) A Pitot tube is mounted on an aircraft flying at sea level in a Standard Atmosphere. The indicator connected to the tube shows a value of 140,548 Pa. What is the flight Mach number of the aircraft?

15) What is the "no slip condition"?

16) Air moves over the hood of your car as you burn down the highway at 150 km/hr. If the car's hood is 1.2 m long, what is the height of the boundary layer when it reaches your windshield (assume the hood is a flat plate)?

17) You and your teammates design a glider for a class project. The geometries of the glider are shown in the following figure.

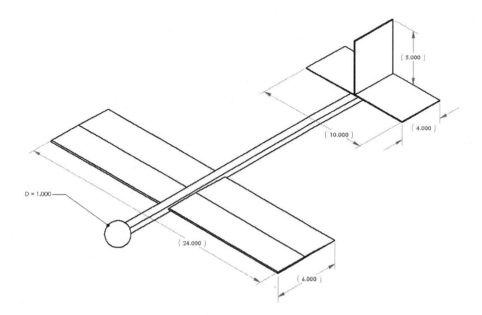

a. If the nose of the glider is made from a 1 inch diameter ball, what is the reference length and area of the nose?

b. If the glider flies in sea level standard air at a speed of 15 ft/sec, what is the Reynolds number with respect to the nose?

c. What is the Reynolds number with respect to the main wing?

d. Use the provided figure to estimate the drag coefficient and the resultant drag generated by the nose.

e. Assuming the flow at the leading edges of the tail surfaces begins as an undisturbed flow, what is the value of the total drag from the tail surfaces? Use flat plate theory for skin friction coefficient to solve.

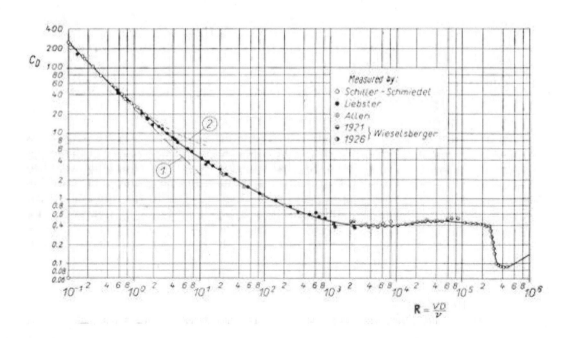

CHAPTER 5 REVIEW PROBLEMS

1) What is the physical difference between airfoil wing sections (2D) and airfoil-based wings (3D)?

2) What is the maximum thickness of a NACA 1410 wing section in terms of percent chord?

3) Where does the maximum deviation between the mean and chord line occur on a NACA 4415 wing section in terms of percent chord?

4) How far does the mean line deviate from the chord line on a NACA 2412 wing section in terms of percent chord?

5) Fill in the following chart for select NACA 4 digit airfoils using the 2D experimental results presented in Appendix C: Sample Airfoil Data:

2-D	NACA 1408 (use Re_c = 3.0 x 10⁶)	NACA 1412 (use Re_c = 3.0 x 10⁶)	NACA 2415 (use Re_c = 9.0 x 10⁶)	NACA 2421 (use Re_c = 5.9 x 10⁶)
$C_{l,max}$				
$\alpha_{L=0}$ $(°)$				
a_0 $(per°)$				
$C_{d,0}$				
$C_{m,\%_4, \alpha=0°}$				

6) Based upon the 2D experimental results from question 5), fill in the chart below for a NACA 2415-based finite wing (Note this is JUST A WING):

3-D Rectangular Wing and Operating Specs			Calculated Values
$c\ (m)$	1	$S_w\ (m^2)$	
$b\ (m)$	8	AR	
$\rho(kg/m^3)$	1.200	$a\ (per°)$	
$V_\infty\ (m/sec)$	119	$C_{D,para}$	
e	0.76	C_L	
$L\ (kN)$	47.5	(L/D)	

CHAPTER 6 REVIEW PROBLEMS

1) The Sopwith Camel has a lift-to-drag ratio of 7.7. Find the glide angle of the famous aircraft in a power-off glide.

2) The stall speed of a fully loaded (W_0 = 140,000 lb$_f$) Boeing 737-800 is 146 kts. The rolling friction coefficient of the landing gears on the runway is 0.025. Its takeoff distance is 6,646 ft. at sea level. The wing area of the aircraft is 1,341 ft². Based on the information provided, find the following:

   a. Estimate the average takeoff force exerted on the aircraft during departure.
   b. What acceleration loading ("g-loading") is exerted on the passengers during takeoff?
   c. If the pair of installed CFM International jet engines are capable of delivering 27,000 lbf thrust each, estimate the coefficient of parasitic drag for the entire aircraft.

3) A small aircraft moving at 156 mph is banked at an angle of 25° traveling at an altitude where the local gravitation acceleration is 31.9 ft/sec². What is the turn radius of the maneuver (in ft.)?

4) A military Unmanned Aerial Vehicle (UAV) is designed with the following specifications:

$$e = 0.87 \qquad S_w = 1.7\,m^2 \qquad b = 4\,m \qquad C_{D,para} = 0.025$$

$$\rho = \rho_{sl,std} \qquad W_0 = 510\,N \qquad W_f = 53.5\,N$$

   a. Find the optimum operating lift coefficient for maximum range
   b. Find the value for optimum range
   c. Find the optimum operating lift coefficient for maximum endurance
   d. Find the value for optimum endurance

5) The British Aerospace Hawk operates at max gross weight of 17,000 pounds, with a turbofan thrust of 5,200 pound thrust. Its wing area is 180 ft² with an aspect ratio AR = 5.3 and e = 0.8. The parasite drag coefficient $C_{D,0}$ = 0.025.

   a. Calculate the thrust required at sea level at max L/D.
   b. From the available excess thrust, calculate the maximum climb rate and angle.
   c. At a thrust specific fuel consumption rate of 1-pound fuel per pound per hour per pound of thrust generated, calculate the gallons of fuel burned per hour in the full throttle climb.
   d. Repeat calculations above at an altitude of 10,000 ft, with T_{max} proportional to air density (varies linearly with density).
   e. Determine the service ceiling altitude at which the climb rate drops to 100 ft/min.

APPENDIX C: SAMPLE AIRFOIL DATA

The following reference pages are provided to the interested reader for the purpose of understanding the fundamental two-dimensional performance specifications of a small but representative selection of general aviation airfoils. The reader is encouraged to seek other resources, in particular the excellent seminal work of NACA researchers Ira H. Abbott, Albert E. Von Doenhoff, and Louis Stivers, Jr., which may be found in NACA records as Technical Report No. 824 [1], or later in a more expansive volume published by Dover [2]. The serious aerospace engineer will find both of these volumes to be irreplaceable assets in aircraft design.

Also included in this section is the presentation of NACA test data of three full scale Clark Y-based wings (4' x 24', 6' x 36', and 8' x 48' wings) tested at Reynolds numbers ranging from 1,000,000 to 9,000,000 [3].

APPENDIX C TABLE OF FIGURES

APPENDIX C TABLE OF TABLES

RECOMMENDED REFERENCES

1. Abbott, Ira H., Albert E. Von Doenhoff, and Louis Stivers Jr. *Summary of airfoil data*. No. NACA-TR-824. 1945.
2. Abbott, Ira H., and A. E. Von Doenhoff. "Theory of Wing Sections: Including a Summary of Airfoil Data (Dover Books on Physics) Author: Ira H. Abbott, A. E. Von Doenhoff." 1959.
3. Silverstein, Abe. "Scale effect on Clark Y airfoil characteristics from NACA full-scale wind-tunnel tests." 1934.

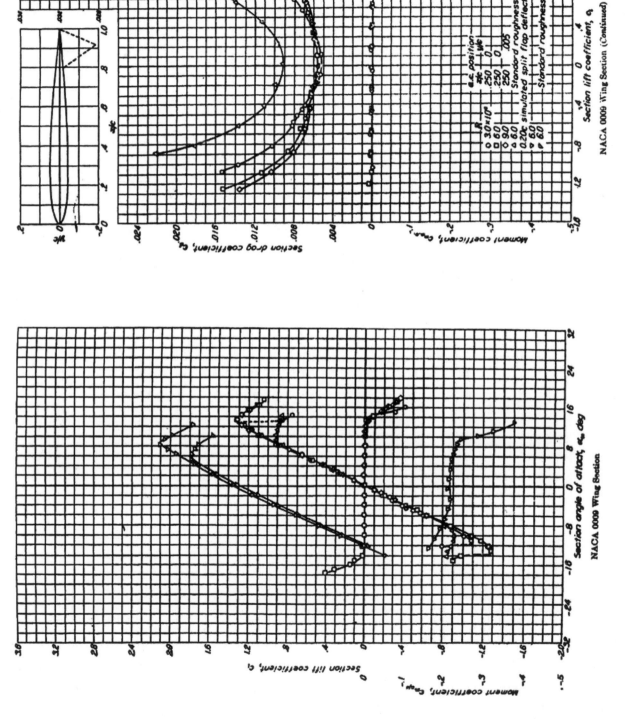

NACA 0009 Wing Section (Continued)

NACA 0009 Wing Section

Figure C-1: NACA 0009 Airfoil Data, $Re_c \in [3.0, 9.0] \times 10^6$

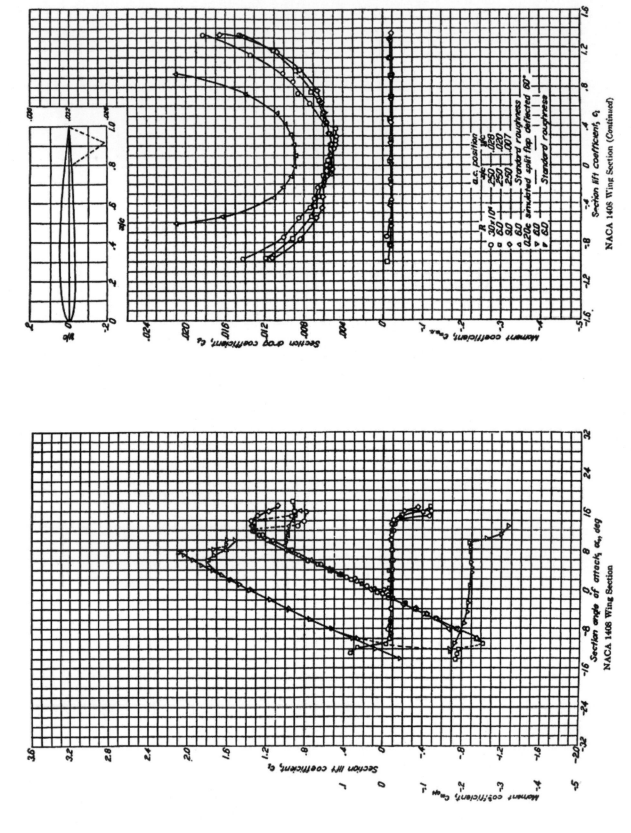

Figure C-2: NACA 1408 Airfoil Data, $Re_c \in [3.0, 9.0] \times 10^6$

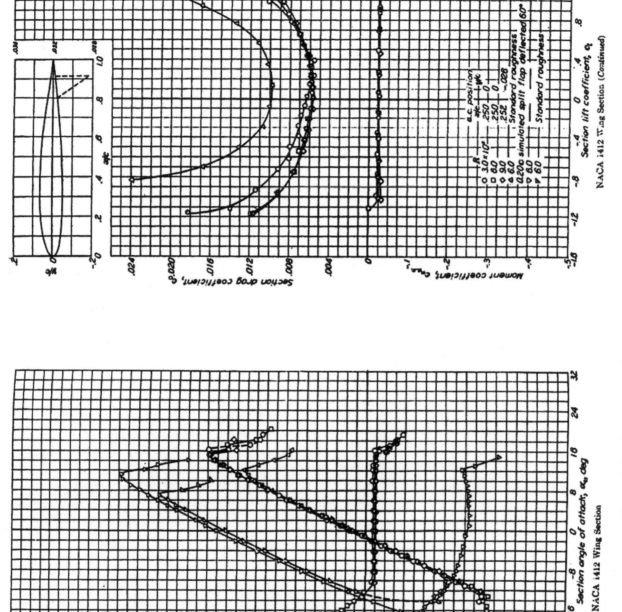

Figure C-3: NACA 1412 Airfoil Data, $Re_c \in [3.0, 9.0] \times 10^6$

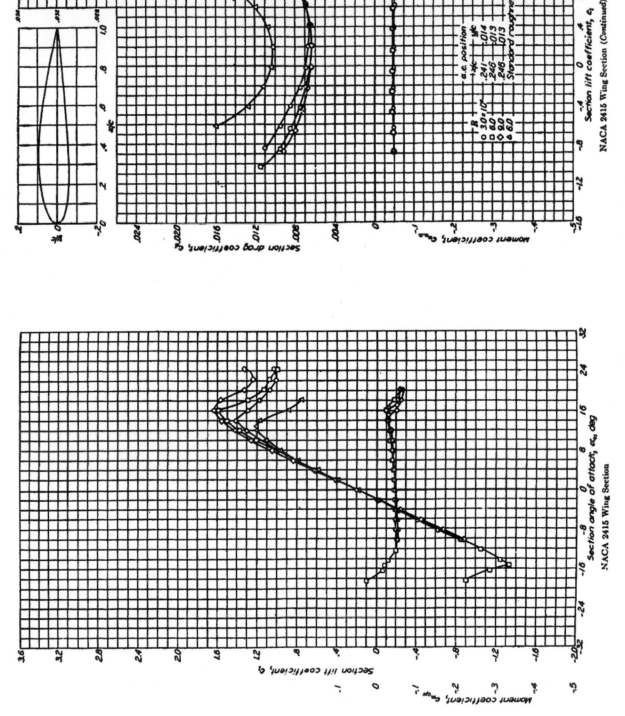

Figure C-4: NACA 2415 Airfoil Data, Re$_c$ ∈ [3.0, 9.0] x 10^6

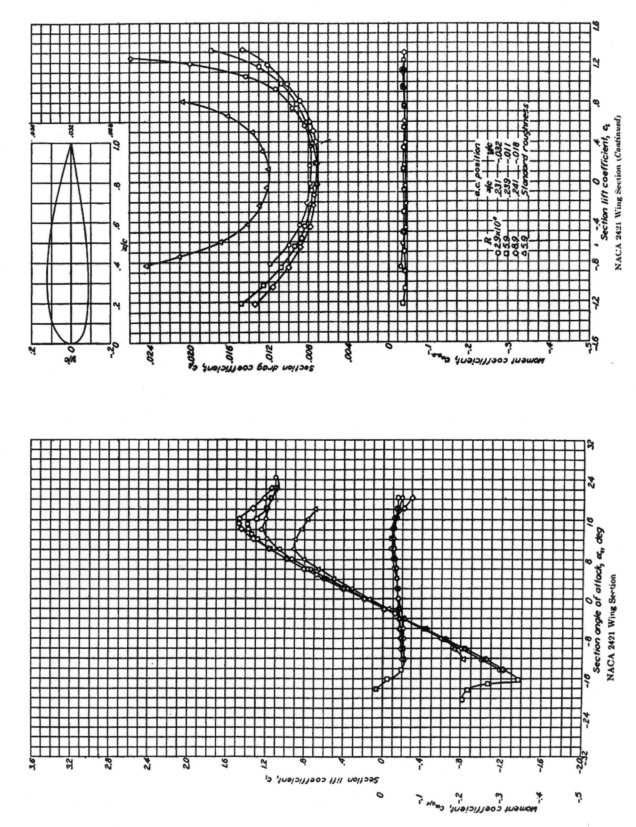

Figure C-5: NACA 2421 Airfoil Data, $Re_c \in [3.0, 9.0] \times 10^6$

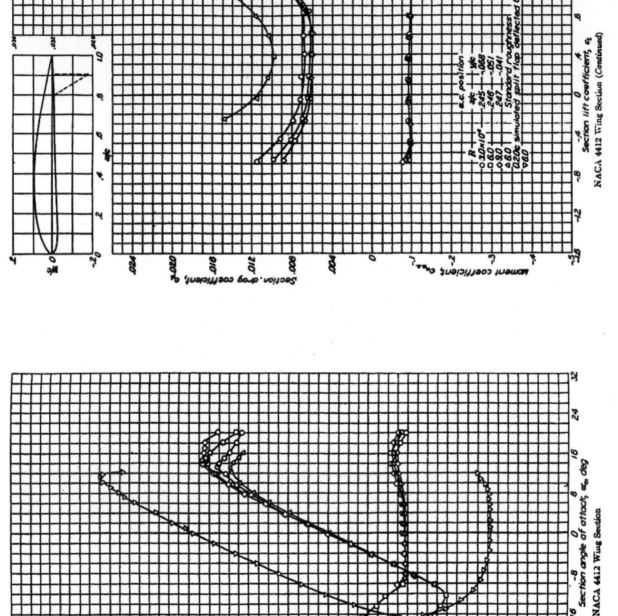

Figure C-6: NACA 4412 Airfoil Data, $Re_c \in [3.0, 9.0] \times 10^6$

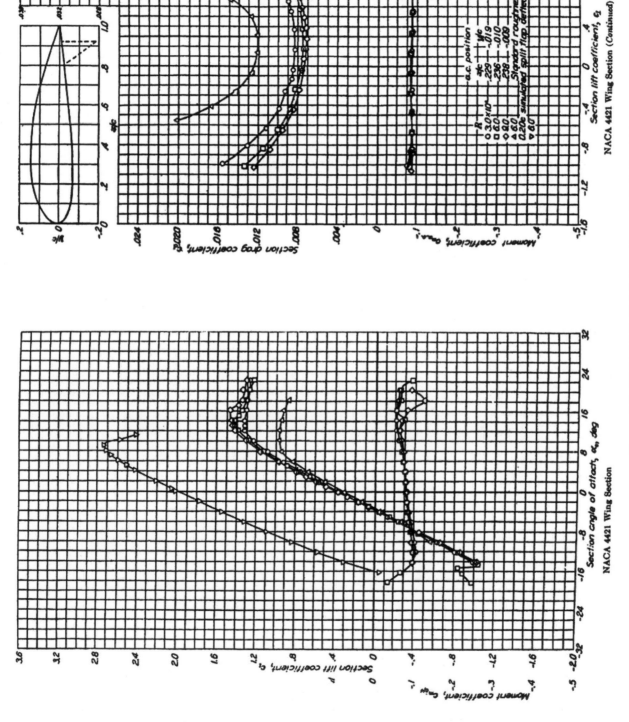

Figure C-7: NACA 4421 Airfoil Data, $Re_c \in [3.0, 9.0] \times 10^6$

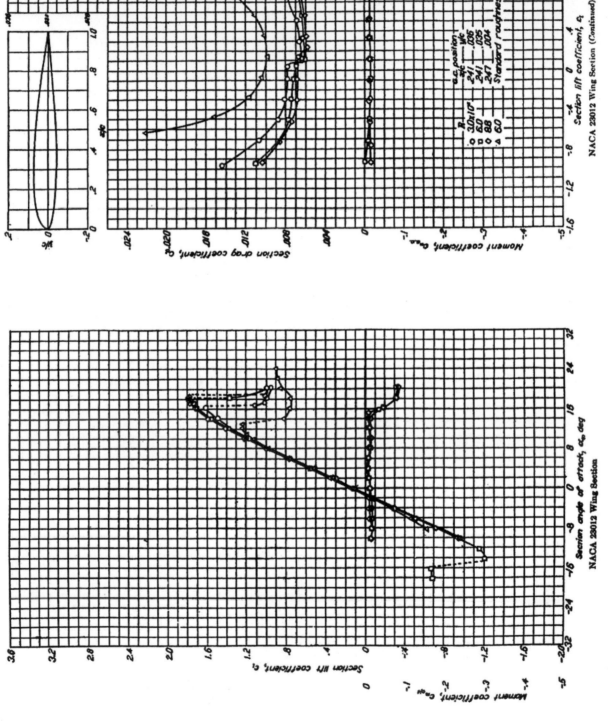

Figure C-8: NACA 23012 Airfoil Data, $Re_c \in [3.0, 9.0] \times 10^6$

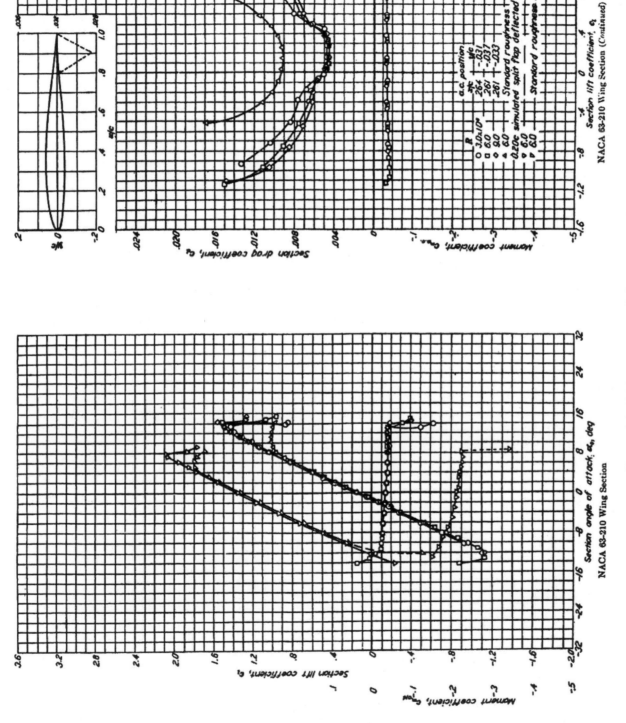

Figure C-9: NACA 63-210 Airfoil Data, $Re_c \in [3.0, 9.0] \times 10^6$

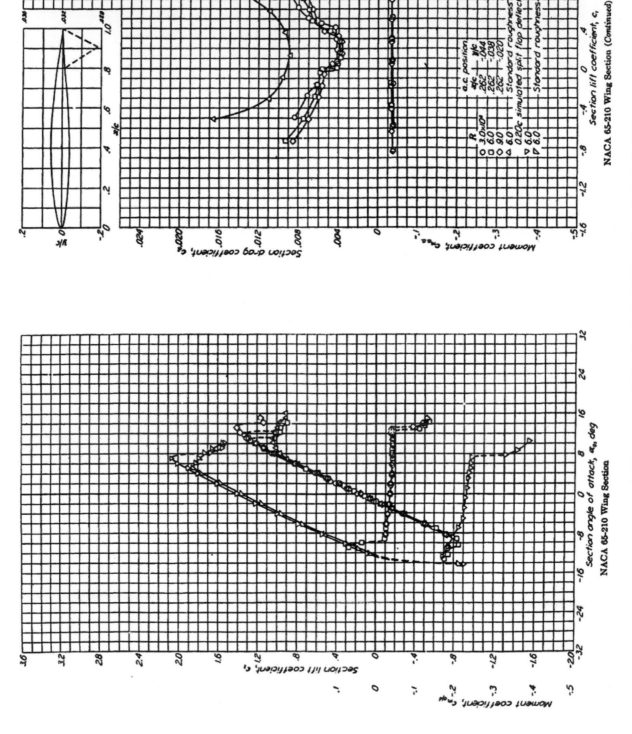

Figure C-10: NACA 65-210 Airfoil Data, $Re_c \in [3.0, 9.0] \times 10^6$

Table C-1: 4' x 24' full scale Clark Y-based wing test results

TABLE II

4 BY 24 CLARK Y AIRFOIL CHARACTERISTICS
R.N.: ZERO LIFT=1.12×10⁶, MAX. LIFT=1.07×10⁶

C_L	α	C_D	L/D	c.p.	$C_{m_c/4}$	C_{D_0}	α_0
−0.2	−9.0	0.0120	−16.7	−11.1	−0.072	0.0098	−8.3
−.1	−7.7	.0100	−10.0	−44.7	−.070	.0094	−7.3
0	−6.2	.0097	0		−.068	.0096	−6.2
.1	−4.8	.0102	9.8	93.8	−.068	.0096	−5.2
.2	−3.3	.0120	16.7	58.7	−.067	.0098	−4.0
.3	−1.9	.0155	19.6	48.0	−.066	.0103	−3.0
.4	−.5	.0200	20.0	41.3	−.065	.0111	−1.9
.5	.0	.0254	19.7	38.0	−.065	.0115	−.9
.6	2.3	.0320	18.8	35.8	−.065	.0120	.2
.7	3.7	.0392	17.9	34.1	−.064	.0119	1.2
.8	5.2	.0493	16.2	32.7	−.062	.0137	2.3
.9	6.7	.0600	15.0	31.5	−.059	.0150	3.5
1.0	8.3	.0738	13.7	30.6	−.056	.0181	4.7
1.1	10.1	.0901	12.2	29.8	−.053	.0237	6.2
1.2	12.0	.1092	11.0	29.2	−.050	.0290	7.7
1.227	13.0	.1240	9.9	30.1	−.062	.0415	8.6
1.2	14.4	.1560	7.7	31.7	−.081	.0788	10.1
1.1	17.2	.2160	5.1	33.3	−.092	.1486	13.3
1.0	20.2	.2759	3.6	34.7	−.100	.2202	16.6
.9	20.7	.3388	2.7	36.7	−.113	.2938	17.5
.8	22.5	.3860	2.1	40.0	−.133	.3504	19.6

TABLE III

4 BY 24 CLARK Y AIRFOIL CHARACTERISTICS
R.N.: ZERO LIFT=1.55×10⁶, MAX. LIFT=1.48×10⁶

C_L	α	C_D	L/D	c.p.	$C_{m_c/4}$	C_{D_0}	α_0
−0.2	−9.0	0.0120	−16.7	−13.1	−0.076	0.0098	−8.4
−.1	−7.7	.0099	−10.1	−50.7	−.076	.0093	−7.3
0	−6.2	.0092	0		−.076	.0088	−6.2
.1	−4.9	.0093	10.8	99.9	−.074	.0087	−5.3
.2	−3.4	.0112	17.9	61.7	−.073	.0090	−4.1
.3	−2.0	.0145	20.7	48.7	−.071	.0095	−3.1
.4	−.6	.0192	20.8	42.0	−.068	.0103	−2.0
.5	.8	.0242	20.6	38.0	−.065	.0103	−1.0
.6	2.2	.0312	19.2	35.6	−.064	.0112	.1
.7	3.6	.0395	17.9	33.8	−.062	.0122	1.1
.8	5.0	.0485	16.5	32.7	−.062	.0129	2.1
.9	6.4	.0582	15.5	31.9	−.062	.0132	3.2
1.0	7.9	.0700	14.3	31.2	−.062	.0143	4.3
1.1	9.6	.0860	12.8	30.7	−.063	.0186	5.7
1.2	11.7	.1093	11.0	30.3	−.064	.0291	7.4
1.255	13.7	.1540	8.1	31.0	−.075	.0664	9.2
1.2	16.4	.2118	5.7	32.8	−.094	.1316	12.1
1.1	19.4	.2660	4.1	34.5	−.107	.1986	15.5
1.0	20.4	.2900	3.5	36.4	−.118	.2343	16.8
.9	20.9	.3039	3.0	38.3	−.128	.2589	17.7

TABLE IV

4 BY 24 CLARK Y AIRFOIL CHARACTERISTICS
R.N.: ZERO LIFT=2.06×10⁶, MAX. LIFT=1.96×10⁶

C_L	α	C_D	L/D	c.p.	$C_{m_c/4}$	C_{D_0}	α_0
−0.2	−8.8	0.0116	−17.2	−13.1	−0.076	0.0094	−8.1
−.1	−7.4	.0095	−10.5	−49.7	−.075	.0089	−7.0
0	−6.0	.0090	0		−.074	.0090	−6.0
.1	−4.6	.0099	10.1	99.8	−.074	.0093	−5.0
.2	−3.2	.0118	16.9	61.7	−.073	.0096	−3.9
.3	−1.8	.0151	19.9	49.0	−.072	.0101	−2.9
.4	−.4	.0197	20.3	42.5	−.070	.0108	−1.8
.5	1.0	.0255	19.6	39.0	−.070	.0116	−.8
.6	2.4	.0325	18.5	36.5	−.069	.0125	.3
.7	3.8	.0404	17.3	34.7	−.068	.0131	1.3
.8	5.0	.0490	16.3	33.4	−.067	.0134	2.4
.9	6.7	.0612	14.7	32.3	−.066	.0162	3.5
1.0	8.2	.0749	13.4	31.5	−.065	.0192	4.6
1.1	9.8	.0901	12.2			.0227	5.9

TABLE V

4 BY 24 CLARK Y AIRFOIL CHARACTERISTICS
R.N.: ZERO LIFT=2.81×10⁶, MAX. LIFT=2.62×10⁶

C_L	α	C_D	L/D	c.p.	$C_{m_c/4}$	C_{D_0}	α_0
−0.2	−8.5	0.0110	−18.3	−13.6	−0.077	0.0088	−7.8
−.1	−7.2	.0098	−10.2	−51.7	−.077	.0090	−6.8
0	−5.8	.0089	0		−.077	.0090	−5.8
.1	−4.4	.0100	10.0	100.8	−.075	.0094	−4.8
.2	−3.0	.0120	16.7	62.2	−.074	.0096	−3.7
.3	−1.6	.0157	19.1	49.0	−.072	.0107	−2.7
.4	−.2	.0201	19.9	42.5	−.070	.0112	−1.6
.5	1.2	.0265	18.9	38.4	−.067	.0126	−.6
.6	2.5	.0330	18.0	36.1	−.067	.0130	.4
.7	3.9	.0411	17.0	34.7	−.068	.0138	1.4
.8	5.3	.0507	15.8	33.5	−.068	.0151	2.4
.9	6.8	.0620	14.5	32.5	−.068	.0170	3.6
1.0	8.2	.0749	13.4	31.5	−.065	.0192	4.8
1.1	9.8	.0901	12.2	30.8	−.064	.0227	5.9
1.2	11.5	.1082	11.1	30.3	−.063	.0250	7.2
1.3	13.1	.1260	10.4	29.9	−.063	.0318	8.5
1.370	14.4	.1405	9.7	29.9	−.067	.0360	9.5
1.3	16.4	.2015	6.5	31.5	−.085	.1073	11.8
1.2	17.4	.2260	5.3	32.2	−.087	.1458	13.1
1.1	20.0	.2788	3.9	33.9	−.100	.2114	16.1
1.0	22.1	.3260	3.1	36.2	−.116	.2703	18.5

TABLE VI

4 BY 24 CLARK Y AIRFOIL CHARACTERISTICS
R.N.: ZERO LIFT=3.19×10⁶, MAX. LIFT=2.96×10⁶

C_L	α	C_D	L/D	c.p.	$C_{m_c/4}$	C_{D_0}	α_0
−0.2	−8.5	0.0100	−20.0	−15.1	−0.080	0.0089	−7.8
−.1	−7.1	.0095	−10.5	−52.7	−.078	.0089	−6.7
0	−5.7	.0089	0		−.077	.0089	−5.7
.1	−4.3	.0096	10.5	100.8	−.075	.0089	−4.7
.2	−2.9	.0118	17.0	62.2	−.074	.0096	−3.6
.3	−1.5	.0155	19.4	49.0	−.072	.0105	−2.6
.4	−.1	.0202	19.8	42.5	−.070	.0113	−1.5
.5	1.3	.0270	18.5	38.6	−.068	.0131	−.5
.6	2.7	.0345	17.4	36.3	−.068	.0145	.6
.7	4.0	.0425	16.5	34.7	−.068	.0152	1.5
.8	5.4	.0525	16.3	33.4	−.067	.0169	2.5
.9	6.9	.0638	14.1	32.4	−.067	.0188	3.7
1.0	8.2	.0755	13.2	31.6	−.066	.0198	4.6
1.1	9.8	.0900	12.2	30.9	−.065	.0226	5.9
1.2	11.3	.1055	11.3	30.0	−.060	.0253	7.0
1.3	13.1	.1261	10.3	29.5	−.058	.0319	8.5
1.381	14.9	.1482	9.3	30.1	−.070	.0419	10.0
1.3	16.0	.1600	8.1	31.0	−.078	.0658	10.4
1.2	18.1	.2400	5.0	32.4	−.090	.1598	13.8
1.1	20.7	.2900	3.8	34.4	−.106	.2226	16.8
1.0	22.7	.3395	2.9	35.5	−.111	.2838	19.1

TABLE VII

4 BY 24 CLARK Y AIRFOIL CHARACTERISTICS
R.N.: ZERO LIFT=3.59×10⁶, MAX. LIFT=3.50×10⁶

C_L	α	C_D	L/D	c.p.	$C_{m_3/4}$	C_{D_0}	α_0
−0.2	−8.4	0.0108	−18.5	−13.0	−0.075	0.0086	−7.7
−.1	−7.0	.0094	−10.6	−49.7	−.075	.0088	−6.6
0	−5.6	.0087	0		−.075	.0089	−5.6
.1	−4.2	.0094	10.6	99.7	−.074	.0088	−4.6
.2	−2.8	.0117	17.1	61.6	−.073	.0090	−3.5
.3	−1.4	.0150	20.0	49.0	−.072	.0100	−2.5
.4	.0	.0199	20.1	42.8	−.071	.0110	−1.4
.5	1.4	.0260	19.2	39.0	−.070	.0121	−.4
.6	2.8	.0338	17.8	36.3	−.068	.0138	.7
.7	4.1	.0420	16.7	34.8	−.067	.0153	1.8
.8	5.6	.0530	15.1	33.2	−.066	.0174	2.7
.9	7.0	.0641	14.1	32.2	−.065	.0191	3.8
1.0	8.4	.0770	13.0	31.4	−.064	.0213	4.8
1.1	10.0	.0920	11.9	30.7	−.063	.0246	6.1
1.2	11.5	.1080	11.1	30.4	−.062	.0273	7.2
1.3	13.1	.1260	10.3	29.6	−.060	.0318	8.5
1.4	14.7	.1461	9.6	29.3	−.060	.0371	9.7
1.420	15.2	.1523	9.3	29.5	−.064	.0406	10.1
1.4	15.5	.1575	8.9	29.8	−.067	.0485	10.5
1.3	18.5	.2434	5.3	32.6	−.100	.1492	13.9
1.2	19.4	.2698	4.4	34.0	−.100	.1896	15.1
1.1	20.8	.2982	3.7	35.2	−.116	.2308	16.9
1.0	23.2	.3355	3.0	36.2	−.118	.2798	19.6

Table C-2: 6' x 36' full scale Clark Y-based wing test results

TABLE VIII
6 BY 36 CLARK Y AIRFOIL CHARACTERISTICS
R.N.: ZERO LIFT=2.07×10⁶, MAX. LIFT=1.90×10⁶

C_L	α	C_D	L/D	^{1}c.p.	$^1C_{m\,c/4}$	C_{D_0}	α_0
−0.1	−7.0	0.0110	−9.0	----	----	0.0105	−6.6
0	−5.7	.0100	0	----	----	.0100	−5.7
.1	−4.3	.0102	9.8	----	----	.0096	−4.7
.2	−2.8	.0117	17.1	----	----	.0091	−3.5
.3	−1.3	.0145	20.7	----	----	.0095	−2.6
.4	−.1	.0189	21.2	----	----	.0092	−1.5
.5	1.3	.0234	21.4	----	----	.0095	−.5
.6	2.7	.0300	20.0	----	----	.0100	.6
.7	4.0	.0382	18.3	----	----	.0109	1.5
.8	5.4	.0476	16.8	----	----	.0120	2.5
.9	6.9	.0591	15.2	----	----	.0134	3.7
1.0	8.3	.0708	14.1	----	----	.0151	4.7
1.1	10.1	.0863	12.7	----	----	.0179	6.0
1.2	11.7	.1020	11.8	----	----	.0218	7.4
1.285	13.6	.1264	10.2	----	----	.0344	9.0
1.2	18.1	.2295	5.3	----	----	.1463	13.8
1.1	19.7	.2606	4.2	----	----	.1932	15.8
1.0	21.5	.3081	3.2	----	----	.2534	17.9

^{1} Not measured.

TABLE IX
6 BY 36 CLARK Y AIRFOIL CHARACTERISTICS
R.N.: ZERO LIFT=3.04×10⁶, MAX. LIFT=2.75×10⁶

C_L	α	C_D	L/D	c.p.	$C_{m\,c/4}$	C_{D_0}	α_0
−0.1	−6.9	0.0096	−10.4	−54.7	−0.080	0.0090	−6.5
0	−5.6	.0088	0	----	−.077	.0088	−5.6
.1	−4.2	.0093	10.8	93.7	−.074	.0087	−4.6
.2	−2.8	.0112	17.8	59.2	−.068	.0090	−3.5
.3	−1.4	.0145	20.7	47.3	−.067	.0095	−2.5
.4	0	.0189	21.2	41.2	−.065	.0100	−1.4
.5	1.4	.0239	20.9	37.8	−.064	.0100	−.4
.6	2.8	.0310	19.4	35.3	−.062	.0110	.7
.7	4.1	.0397	18.1	33.7	−.061	.0114	1.6
.8	5.6	.0476	16.8	32.2	−.058	.0120	2.7
.9	7.0	.0585	15.4	31.3	−.057	.0130	3.8
1.0	8.3	.0707	14.3	30.6	−.056	.0145	4.7
1.1	10.0	.0843	13.0	29.8	−.053	.0163	6.1
1.2	11.5	.0990	12.1	29.2	−.060	.0183	7.2
1.3	13.3	.1180	11.0	23.8	−.049	.0238	8.7
1.330	14.2	.1307	10.2	28.9	−.052	.0322	9.5
1.3	14.7	.1485	8.8	30.0	−.065	.0543	10.1
1.2	18.9	.2440	4.9	32.2	−.087	.1638	14.6
1.1	20.7	.2828	3.9	34.2	−.104	.2154	16.8
1.0	22.0	.3295	3.0	36.2	−.117	.2638	18.4

TABLE X
6 BY 36 CLARK Y AIRFOIL CHARACTERISTICS
R.N.: ZERO LIFT=3.64×10⁶, MAX. LIFT=3.22×10⁶

C_L	α	C_D	L/D	c.p.	$C_{m\,c/4}$	C_{D_0}	α_0
−0.1	−6.9	0.0101	−9.9	−57.6	−0.083	0.0095	−6.5
0	−5.5	.0090	0	----	−.080	.0090	−5.5
.1	−4.2	.0095	10.5	103.8	−.078	.0089	−4.6
.2	−2.7	.0112	17.9	62.6	−.075	.0090	−3.4
.3	−1.3	.0145	20.7	48.7	−.071	.0095	−2.4
.4	.1	.0194	20.6	42.0	−.068	.0093	−1.3
.5	1.5	.0246	20.3	38.0	−.065	.0107	−.3
.6	2.8	.0312	19.2	35.5	−.063	.0112	.7
.7	4.2	.0396	17.7	33.7	−.061	.0123	1.7
.8	5.7	.0491	16.3	32.5	−.060	.0134	2.8
.9	7.0	.0585	15.4	31.7	−.060	.0140	3.8
1.0	8.4	.0708	14.1	31.0	−.060	.0147	4.8
1.1	9.9	.0828	13.3	30.5	−.060	.0154	6.0
1.2	11.4	.0978	12.3	30.0	−.060	.0176	7.1
1.3	13.0	.1155	11.3	29.3	−.056	.0213	8.4
1.36	14.7	.1383	9.8	29.0	−.055	.0353	9.8
1.3	15.0	.1555	8.4	29.7	−.061	.0613	10.4
1.2	19.2	.2545	4.7	33.4	−.102	.1743	14.9
1.1	21.6	.3033	3.6	35.2	−.115	.2359	17.7

TABLE XI
6 BY 36 CLARK Y AIRFOIL CHARACTERISTICS
R.N.: ZERO LIFT=4.15×10⁶, MAX. LIFT=3.64×10⁶

C_L	α	C_D	L/D	c.p.	$C_{m\,c/4}$	C_{D_0}	α_0
−0.2	−8.4						−7.7
−.1	−6.9	0.0098	−10.2	−58.6	−0.084	0.0092	−6.5
0	−5.6	.0088	0	----	−.080	.0088	−5.6
.1	−4.2	.0094	10.6	102.8	−.077	.0088	−4.6
.2	−2.7	.0111	18.0	62.6	−.075	.0089	−3.4
.3	−1.3	.0145	20.7	49.4	−.073	.0095	−2.4
.4	.1	.0191	20.9	42.2	−.069	.0102	−1.3
.5	1.5	.0243	20.6	38.2	−.066	.0104	−.3
.6	2.9	.0313	19.2	35.5	−.063	.0113	.8
.7	4.2	.0388	18.1	33.8	−.062	.0115	1.7
.8	5.7	.0478	16.7	32.7	−.062	.0122	2.8
.9	7.1	.0585	14.3	31.8	−.061	.0136	3.9
1.0	8.5	.0701	14.3	31.0	−.060	.0144	4.9
1.1	10.1	.0841	13.1	30.3	−.058	.0160	6.2
1.2	11.6	.0989	12.1	29.8	−.057	.0187	7.3
1.3	13.3	.1167	11.1	29.3	−.055	.0225	8.7
1.371	15.0	.1382	9.9	30.4	−.073	.0334	10.1
1.3	15.6	.1695	7.7	29.8	−.062	.0753	11.0
1.2	18.5	.2338	5.1	32.4	−.090	.1536	14.2
1.1	21.2	.2853	3.8	34.1	−.102	.2184	17.3

TABLE XII
6 BY 36 CLARK Y AIRFOIL CHARACTERISTICS
R.N.: ZERO LIFT=4.77×10⁶, MAX. LIFT=4.20×10⁶

C_L	α	C_D	L/D	c.p.	$C_{m\,c/4}$	C_{D_0}	α_0
−0.2	−8.2	0.0122	−16.4	−18.6	−0.087	0.0100	−7.5
−.1	−6.8	.0101	−9.9	−54.6	−.080	.0095	−6.4
0	−5.6	.0089	0	----	−.076	.0089	−5.6
.1	−4.1	.0093	10.7	98.7	−.073	.0087	−4.5
.2	−2.7	.0110	18.2	60.1	−.070	.0088	−3.4
.3	−1.4	.0140	21.4	47.4	−.067	.0090	−2.5
.4	−.1	.0188	21.5	41.3	−.065	.0097	−1.5
.5	1.4	.0234	21.4	37.8	−.064	.0095	−.4
.6	2.8	.0300	20.0	35.6	−.063	.0100	.5
.7	3.9	.0376	18.6	33.7	−.061	.0103	1.4
.8	5.4	.0475	16.8	32.5	−.060	.0119	2.6
.9	6.7	.0570	15.8	31.7	−.060	.0120	3.6
1.0	8.0	.0676	14.8	31.0	−.060	.0119	4.4
1.1	9.4	.0798	13.8	30.3	−.058	.0124	5.5
1.2	10.8	.0935	12.9	29.7	−.056	.0133	6.5
1.3	12.4	.1096	11.9	29.1	−.053	.0154	7.8
1.4	14.3	.1299	10.8	28.7	−.052	.0209	9.3
1.448	15.6	.1483	9.8	29.3	−.062	.0313	10.4
1.4	15.8	.1530	9.2	29.8	−.007	.0440	10.8
1.3	16.0	.1815	7.2	30.3	−.069	.0873	11.4
1.2	19.6	.2675	4.5	33.1	−.099	.2593	15.3
1.1	22.8	.3396	3.2	34.3	−.106	.2722	18.9

TABLE XIII
6 BY 36 CLARK Y AIRFOIL CHARACTERISTICS
R.N.: ZERO LIFT=5.86×10⁶

C_L	α	C_D	L/D	c.p.	$C_{m\,c/4}$	C_{D_0}	α_0
−0.1	−6.9	0.0097	10.3	−56.5	−0.032	0.0091	−6.5
0	−5.5	.0090	0	----	−.077	.0090	−5.5
.1	−4.4	.0094	10.6	98.7	−.073	.0088	−4.5
.2	−2.6	.0112	17.9	60.1	−.070	.0090	−3.3
.3	−1.2	.0147	20.4	47.7	−.068	.0097	−2.2
.4	.2	.0200	20.0	41.8	−.067	.0111	−1.2
.5	1.7	.0254	19.7	38.0	−.065	.0115	−.1
.6	3.1	.0322	18.6	35.6	−.064	.0122	1.0
.7	4.3	.0402	17.4	34.0	−.063	.0129	2.0
.8	5.8	.0496	16.2	32.9	−.063	.0140	3.0

Table C-3: 8' x 48' full scale Clark Y-based wing test results

TABLE XIV

8 BY 48 CLARK Y AIRFOIL CHARACTERISTICS
R.N.: ZERO LIFT=2.20×10⁶, MAX. LIFT=1.84×10⁶

C_L	α	C_D	L/D	c.p.	$C_{m_{c/4}}$	C_{D_0}	α_0
−0.2	−8.4	0.0131	−15.3	−15.6	−0.031	0.0109	−7.7
−.1	−7.0	.0108	−9.3	−52.5	−.078	.0103	−6.6
0	−5.6	.0090	0		−.075	.0099	−5.6
.1	−4.2	.0099	10.1	98.7	−.073	.0093	−4.5
.2	−2.8	.0110	18.2	60.6	−.071	.0088	−3.6
.3	−1.4	.0134	22.4	48.4	−.070	.0084	−2.5
.4	0	.0175	22.9	42.5	−.070	.0086	−1.4
.5	1.4	.0230	21.7	38.8	−.069	.0091	−.4
.6	2.7	.0305	19.7	36.3	−.068	.0104	.5
.7	4.1	.0387	18.1	34.4	−.066	.0114	1.6
.8	5.5	.0483	16.6	33.1	−.065	.0127	2.7
.9	7.0	.0590	15.3	32.2	−.065	.0139	3.8
1.0	8.6	.0721	13.9	31.5	−.065	.0164	5.0
1.1	10.2	.0876	12.6	30.7	−.063	.0202	6.3
1.2	11.9	.1040	11.5	30.2	−.063	.0238	7.6
1.3	13.7	.1233	10.5	29.8	−.062	.0291	9.0
1.325	14.7	.1399	9.5	30.0	−.066	.0420	10.0
1.3	15.8	.1644	7.9	31.0	−.078	.0703	11.2
1.2	17.7	.2133	5.6	33.0	−.096	.1331	13.4
1.1	19.6	.2591	4.2	35.0	−.112	.1917	15.6
1.0	21.9	.3191	3.1	37.0	−.125	.2635	18.3

TABLE XV

8 BY 48 CLARK Y AIRFOIL CHARACTERISTICS
R.N.: ZERO LIFT=3.10×10⁶, MAX. LIFT=2.59×10⁶

C_L	α	C_D	L/D	c.p.	$C_{m_{c/4}}$	C_{D_0}	α_0
−0.2	−8.3	0.0130	−15.4	−17.0	−0.084	0.0108	−7.6
−.1	−7.0	.0101	−9.9	−54.6	−.080	.0095	−6.6
0	−5.5	.0091	0		−.076	.0091	−5.5
.1	−4.1	.0093	10.8	98.7	−.073	.0087	−4.5
.2	−2.7	.0106	18.0	60.1	−.070	.0084	−3.4
.3	−1.3	.0131	22.9	48.0	−.009	.0081	−2.4
.4	.1	.0180	22.2	41.8	−.067	.0091	−1.3
.5	1.5	.0234	21.4	38.4	−.067	.0095	−.4
.6	2.8	.0305	19.7	36.2	−.067	.0104	.7
.7	4.2	.0389	18.0	34.5	−.067	.0116	1.7
.8	5.6	.0481	16.6	33.4	−.067	.0125	2.7
.9	7.0	.0590	15.3	32.4	−.067	.0138	3.8
1.0	8.6	.0721	13.8	31.7	−.067	.0164	5.0
1.1	10.1	.0870	12.7	31.2	−.068	.0196	6.2
1.2	11.8	.1046	11.5	30.6	−.067	.0244	7.5
1.3	13.6	.1228	10.6	30.2	−.067	.0311	8.9
1.35	15.2	.1442	9.4	29.9	−.066	.0412	10.3
1.3	16.4	.1772	7.3	31.9	−.090	.0831	11.8
1.2	18.4	.2333	5.1	33.8	−.106	.1531	14.1
1.1	20.6	.2811	3.9	35.3	−.116	.2137	16.6
1.0	22.8	.3256	3.1	36.5	−.120	.3597	19.3

TABLE XVI

8 BY 48 CLARK Y AIRFOIL CHARACTERISTICS
R.N.: ZERO LIFT=4.13×10⁶, MAX. LIFT=3.78×10⁶

C_L	α	C_D	L/D	c.p.	$C_{m_{c/4}}$	C_{D_0}	α_0
−0.2	−8.2	0.0126	−15.9	−17.1	−0.084	0.0104	−7.5
−.1	−6.9	.0100	−10.0	−53.6	−.079	.0094	−6.5
0	−5.5	.0090	0		−.075	.0090	−5.5
.1	−4.1	.0095	10.5	98.7	−.073	.0089	−4.5
.2	−2.8	.0109	18.4	60.6	−.071	.0087	−3.5
.3	−1.4	.0137	21.9	48.4	−.070	.0087	−2.5
.4	0	.0175	22.9	42.5	−.070	.0086	−1.4
.5	1.3	.0228	21.9	39.0	−.070	.0089	−.5
.6	2.6	.0295	20.3	36.5	−.069	.0094	.5
.7	4.0	.0377	18.6	34.8	−.069	.0104	1.6
.8	5.3	.0465	17.2	33.5	−.068	.0109	2.6
.9	6.7	.0573	15.7	32.6	−.068	.0122	3.6
1.0	8.2	.0690	14.5	31.8	−.068	.0133	4.6
1.1	9.6	.0824	13.3	31.1	−.067	.0150	5.7
1.2	11.2	.0998	12.0	30.5	−.066	.0196	6.9
1.3	12.8	.1182	10.9	30.0	−.065	.0241	8.2
1.4	14.7	.1398	10.0	29.7	−.065	.0306	9.7
1.445	15.9	.1550	9.3	30.2	−.075	.0387	10.7
1.4	17.4	.2058	6.8	30.4	−.075	.0966	12.4
1.3	19.0	.2492	5.2	32.3	−.095	.1551	14.4
1.2	21.6	.2832	4.2	34.8	−.120	.2304	17.3

TABLE XVII

8 BY 48 CLARK Y AIRFOIL CHARACTERISTICS
R.N.: ZERO LIFT=5.58×10⁶, MAX. LIFT=4.43×10⁶

C_L	α	C_D	L/D	c.p.	$C_{m_{c/4}}$	C_{D_0}	α_0
−0.2	−8.1	0.0119	−16.8	−15.1	−0.080	0.0097	−7.4
−.1	−6.8	.0098	−10.2	−51.7	−.077	.0092	−6.4
.0	−5.4	.0088	0		−.076	.0088	−5.4
.1	−4.0	.0091	11.0	100.6	−.075	.0085	−4.4
.2	−2.6	.0106	18.9	62.6	−.075	.0084	−3.3
.3	−1.2	.0137	21.9	49.7	−.074	.0087	−2.3
.4	.2	.0183	21.9	43.5	−.074	.0094	−1.3
.5	1.5	.0234	21.4	39.6	−.073	.0095	−.3
.6	2.8	.0303	19.8	36.8	−.071	.0102	.7
.7	4.2	.0282	18.3	35.0	−.070	.0109	1.7
.8	5.5	.0465	17.2	33.6	−.069	.0116	2.6
.9	6.9	.0578	15.6	32.4	−.067	.0127	3.7
1.0	8.3	.0696	14.3	31.5	−.065	.0139	4.7
1.1	9.8	.0831	13.2	31.0	−.066	.0162	5.9
1.2	11.4	.1003	12.0	30.6	−.067	.0201	7.1
1.3	13.0	.1182	11.0	30.2	−.067	.0241	8.4
1.4	14.8	.1388	10.1	30.0	−.069	.0296	9.8
1.46	16.2	.1600	9.1	29.7	−.068	.0413	11.0
1.4	16.9	.1763	7.9	30.0	−.089	.0676	11.9
1.3	17.7	.2190	5.9	31.9	−.090	.1251	13.1
1.2	22.1	.3210	3.7	35.2	−.125	.2406	17.8

TABLE XVIII

8 BY 48 CLARK Y AIRFOIL CHARACTERISTICS
R.N.: ZERO LIFT=6.12×10⁶, MAX. LIFT=5.38×10⁶

C_L	α	C_D	L/D	c.p.	$C_{m_{c/4}}$	C_{D_0}	α_0
−0.2	−8.1	0.0124	−16.1	−15.0	−0.080	0.0102	−7.4
−.1	−6.7	.0097	−10.3	−54.0	−.078	.0091	−6.3
.0	−5.3	.0086	0		−.076	.0086	−5.3
.1	−3.9	.0089	10.2	96.1	−.075	.0083	−4.3
.2	−2.6	.0106	18.9	61.9	−.073	.0084	−3.3
.3	−1.2	.0139	21.6	49.5	−.072	.0089	−2.3
.4	.2	.0181	22.1	43.0	−.072	.0092	−1.3
.5	1.5	.0234	21.4	39.5	−.071	.0095	−.3
.6	2.8	.0305	19.7	37.0	−.071	.0104	.7
.7	4.2	.0382	18.3	35.0	−.071	.0109	1.7
.8	5.6	.0476	16.8	34.0	−.071	.0120	2.7
.9	6.9	.0581	15.5	33.0	−.070	.0130	3.7
1.0	8.4	.0696	14.4	32.1	−.070	.0139	4.8
1.1	9.8	.0836	13.2	31.3	−.070	.0162	5.9
1.2	11.3	.0999	12.0	30.9	−.070	.0186	7.0
1.3	12.9	.1170	11.0	30.3	−.070	.0231	8.3
1.4	14.7	.1380	10.1	30.0	−.070	.0286	9.7
1.51	16.7	.1660	9.1	29.9	−.070	.0390	11.3
1.4	17.3	.1910	7.3	29.8	−.066	.0816	12.3
1.3	19.2	.2570	5.1	33.6	−.110	.1631	14.6
1.2	22.0	.3150	3.8	34.9	−.118	.2346	17.7

TABLE XIX

8 BY 48 CLARK Y AIRFOIL CHARACTERISTICS
R.N.: ZERO LIFT=7.53×10⁶

C_L	α	C_D	L/D	c.p.	$C_{m_{c/4}}$	C_{D_0}	α_0
−0.2	−8.3	0.0126	−15.9	−15.1	−0.080	0.0104	−7.6
−.1	−6.8	.0097	−10.3	−51.7	−.077	.0091	−6.4
.0	−5.3	.0086	0		−.075	.0086	−6.3
.1	−3.8	.0092	10.9	99.6	−.074	.0086	−4.2
.2	−2.3	.0112	17.9	61.6	−.073	.0090	−3.0
.3	−1.0	.0147	20.4	48.7	−.071	.0097	−2.1
.4	.3	.0183	21.8	42.5	−.070	.0094	−1.2
.5	1.6	.0240	20.8	39.0	−.070	.0101	−.2
.6	2.9	.0305	19.7	36.7	−.070	.0104	.8
.7	4.3	.0385	18.2	35.0	−.070	.0112	1.8
.8	5.6	.0464	17.3	33.9	−.071	.0108	2.7

TABLE XX

8 BY 48 CLARK Y AIRFOIL CHARACTERISTICS
R.N.: ZERO LIFT=8.77×10⁶

C_L	α	C_D	L/D	c.p.	$C_{m_{c/4}}$	C_{D_0}	α_0
−0.1	−7.2	0.0095	−10.5	−55.7	−0.081	0.0089	−6.8
.0	−5.4	.0087	0		−.076	.0037	−5.4
.1	−3.7	.0091	11.0	99.6	−.074	.0085	−4.1
.2	−2.3	.0108	18.5	61.6	−.073	.0086	−3.0
.3	−1.0	.0139	21.6	49.0	−.072	.0089	−2.1
.4	.2	.0180	22.2	42.8	−.070	.0091	−1.2
.5	1.5	.0229	21.8	39.2	−.071	.0090	−.3
.6	2.8	.0285	21.6	37.0	−.072	.0084	.6

APPENDIX D: ENGINEERING PROBLEM SOLVING

One of the most fundamentally important skills of any engineer is that of general problem solving. An organized approach to any problem can often yield the most effective solution with the least amount of effort. While there are no doubt many methods which one may adopt, the basic approach presented here has been used by the author for many years. The following infographic highlights the basic engineering problem solving method:

Problem Description (*Identification* Process)

- State the problem precisely
- Draw a sketch or diagram of the problem
- List the given variables

Mission Development (*Decision* Process)

- Determine the required variable(s) in desired units
- List relevant equations, constants, assumptions made, and additional notes

Solution Procedure (*Execution* Process)

- Carefully document the solution steps
- State the final answer with appropriate units

Result Evaluation (*Assessment* Process)

- Has the answer appropriately addressed the initial problem?
- Do all aspects of the answer (magnitude, sign/direction, units, etc.) make sense?

In recent years, the field of engineering has benefited immensely from the introduction of the World Wide Web. Since its inception in the early 1990's, the internet has provided researchers, scientists and engineers with a huge (and rapidly growing) information database of information. Of course, all information is not created equal, and much of what is available online is complete bunk. Thus, it behooves the careful engineer to develop his or her own experience-based "engineering sense" to avoid falling victim to misinformation. As it relates to the proposed problem solving method herein disclosed, the author's own experience has shown that today's newer engineers have an incredible gift in both the DECISION and EXECUTION phases of the process. On the other hand, these same engineers are often severely handicapped in their ability to properly IDENTIFY the true problem posed, and even less adept in ASSESSMENT of their own results. Fortunately, both of these deficiencies can be overcome through diligent practice in the "art" of engineering.

The proposed 4-step process (IDENTIFY-DECIDE-EXECUTE-ASSESS, or "IDEA") can be applied to nearly any engineering problem and is demonstrated in the following example.

Problem Statement (IDENTIFY):

A purple box with ½ inch thick walls and internal measurements of 12 x 12 x 8 inches is filled with water at 72.5 °F. How many gallons of water can the box hold?

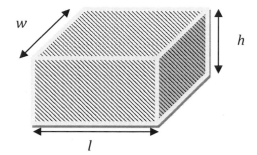

$$t = 0.5\,in \qquad T = 72.5\,°F$$
$$l = 12\,in \qquad w = 12\,in \qquad h = 8\,in$$

Find (DECIDE):

$$\Psi = ?\,US\,gals$$

Relevant Equations, List of Constants, Assumptions and Other Notes (DECIDE):

$$\Psi = l \times w \times h \qquad\qquad \Psi \neq f(T, t, color)$$
$$1\,US\,gal = 0.1337\,ft^3 \qquad\qquad 1\,ft = 12\,in$$

1. The thickness of the walls is an unimportant known value since the internal dimensions govern the containment volume of the box.
2. The color of the box is an unimportant known value since it does not impact its volume.
3. The temperature of the water is only important in the sense that it signifies that the water is likely not boiling or freezing, based on Assumption #1.
4. The internal dimensions are important known values since they dictate the volume of the box, based upon Assumption #2.
5. Assumption #1: The ambient pressure is that of a Standard Atmosphere at sea level.
6. Assumption #2: The box is completely rigid.
7. Assumption #3: Aside from the top, there are no holes in the box.

Solution (EXECUTE):

$$\Psi = (12\,in)(12\,in)(8\,in)\left(\frac{1\,US\,gal}{0.1337\,ft^3}\right)\left(\frac{1\,ft}{12\,inches}\right)^3 = 1152\,in^3\left(\frac{1\,US\,gal}{0.1337\,ft^3}\right)\left(\frac{1\,ft^3}{1728\,in^3}\right);$$

$$\boxed{\Psi = 4.99\,US\,gals}$$

Check (ASSESS):

The answer (5 US gallons) directly addresses the initial problem statement, uses the correct units, and seems to be an appropriate magnitude given the dimensions of the box – the answer checks.

INDEX

A

A-6A (aircraft), viii, 152
Abbott, Ira, 131, 202
adiabatic, 89, 90, 91, 93, 94, 118
Aerodrome (aircraft), 9
altitude
  absolute, 40, 48
  density, 49, 66
  geometric, 40, 41, 48, 49, 50, 51, 54, 56, 58, 61, 66, 196, 198
  geopotential, 48, 50, 56, 57, 61, 62, 66, 196
  pressure, 49, 50, 66
  temperature, 66
angle
  glide, ix, 179, 193, 201
  of attack, ix, 32, 101, 108, 126, 129, 130, 136, 141, 142, 146, 151, 156, 159, 160, 163, 175, 193
  of incidence, 175
  zero lift, 130, 156
Archimedes
  of Syracuse, 28
  principle, 28
argon, 16, 55
aspect ratio, xi, 139, 140, 160, 161, 163, 164, 191, 194, 195, 201
astronomical units, 44
attitude, 151, 165, 171, 178

B

Bell X-1 (aircraft), 165
Bernoulli
  Daniel, 107, 124
  equation, 81, 118, 121, 198
Boeing
  737-800 (aircraft), 201
  747 (aircraft), 1, 2, 10, 146, 148, 167, 191
  777 (aircraft), 9
Boltzmann constant, ix, 20
boundary layer, ix, xi, 67, 70, 103, 104, 105, 108, 119, 122, 148, 150, 153, 163, 198
  laminar, 153
  transitional, x, 100, 104
  turbulent, 154
Brahe, Tycho, 41
Buckingham
  Edgar, 96
  pi theorem, 96

C

carbon dioxide, 55
Carnot
  cycle, 166
  Nicolas, 166
Cayley, Sir George, 6, 7, 123
ceiling
  absolute, 187, 188
  service, 180, 187, 188, 201
Celsius
  (degrees), 12, 14
  Anders, 20
center of mass, 38
Cessna
  150 (aircraft), viii, 1, 3, 4
  206-U (aircarft), viii
circle (conic section), 42
circulation, ix, 127, 148, 151, 159, 163, 195
  control, viii, xi, 151
Clark Y, 125, 192, 202, 213, 214, 215
coefficient of heat
  at constant pressure, 86
  at constant volume, 84, 86
conic sections, 42
conservation
  of energy, 11, 88
  of mass, 11, 77
  of momentum, 11
continuity. *See* conservation of mass
control surface (for a fluid body), xi, 3, 11, 75, 79
control surfaces (for aircarft control), 3, 5, 9, 12, 123
control volume, xi, 75, 77, 92, 163

D

da Vinci, Leonardo, 5, 124
d'Alembert
  Jean le Rond, 107
  paradox, 107
density
  air, 11, 25, 36, 46, 66, 74, 82, 172, 177, 180, 187, 198, 201
  energy, 180, 222
  water, 35, 196
dew point, 48
dihedral, 195
drag

coefficient, 98, 102, 106, 112, 115, 119, 122, 130, 142, 152, 153, 160, 165, 166, 199

    induced, 108, 109, 136, 137, 139, 140, 142, 144, 145, 149, 160, 161, 162, 177

    interference, 108

    parasite, 108, 132, 142, 145, 184, 185, 187, 191, 201

    pressure, 68, 108

    skin friction, 68, 108, 119

    wave, 108, 165, 166

Drela, Mark, 154

Du Temple, Felix, 8

E

efficiency

    Carnot, 167

    propeller, 173, 175

    spanwise load. *See* Oswald efficiency

    thermal, 166, 173

ellipse (conic section), 42

energy

    balance, 89, 90, 91

    equation, 88, 90, 118

    heat, 88

    internal, ix, 24, 67, 82, 83, 84, 86, 87, 90

    kinetic, xi, 20, 30, 31, 34, 36, 38, 71, 82, 87, 90, 138, 173, 175, 177

    potential, xi, 38

    rotational, 83, 84

    translational, 83

    vibrational, 83

    work, 88

Euler, Leonhart, 107

Eulerian, 69, 70

F

Fahrenheit

    (degrees), 12, 14

    Daniel, 20

flat plate, 100, 101, 103, 104, 105, 106, 119, 122, 153, 163, 198, 199

Ford 4-AT Trimotor (aircraft), 9

G

Galilei, Galileo, 20

Gibbs-Smith, C. H., 8

Gossamer Condor (aircraft), 5, 32

Göttingen 398, 125

gravity, 13, 40, 47, 56, 57, 58, 59, 60, 66, 78, 104, 118, 179, 181, 196

    center of, 1, 4, 10, 12, 123, 192, 193, 194

    specific, xi, 17, 18, 19, 29, 35, 46, 195

H

Hawk (aircraft), 9, 201

helium, 16, 31, 36, 55, 83, 196

Henson, William S., 6

Hepperle, Martin, 154

hydrogen, 9, 55, 66

hyperbola (conic section), 42

I

ideal gas, 24, 34, 36, 54, 67, 68

impulse function, 78, 79, 80, 121

isentropic, 48, 87, 91, 94, 95, 108, 118, 122, 198

    relations, 94, 95, 118

J

JavaFoil, 154, 155, 156, 157

Joukowski, Nikolai, 127

K

Kelvin

    (measure of temperature), 12, 14, 20

    Lord, 20

Kepler

    Johannes, 41

    laws of planetary motion, 41, 45

krypton, 55

Kutta

    condition, 127, 128, 153, 195

    -Joukowski theorem, 127

    Wilhelm, 127

L

Lagrangian, 69

Lanchester

    Frederick, 127

Langley, Samuel P., 9

lateral axis, 171

Liebeck

    airfoil, 32

    Robert, 32, 125

lift

    coefficient, 125, 129, 130, 141, 142, 145, 146, 148, 151, 152, 160, 161, 162, 164, 167, 170, 172, 183, 184, 195, 201

Lilienthal, Otto, 8, 9, 124

Lissaman, Peter, 32

longitudinal axis, 171

Low Earth Orbit, 40

U

unsteady flow, 68, 69

V

velocity
    cruise, 144, 145, 147, 162, 165, 172
    escape, 38, 39
    freestream, 81, 103, 108, 111, 112, 126, 136, 163, 174
    landing, 3, 186
    orbital, 9, 40, 41, 45, 64, 66, 196
    stall, 146, 147, 148, 162, 164, 184
    takeoff, 185
vertical axis, 21, 171, 179
viscosity, ix, 67, 100, 101, 105, 108, 122
von Doenhoff, Albert, 131, 202
vortex, ix, 31, 127, 136, 148, 153, 154, 158, 159, 163
    lattice method, 158
    panel method, 153, 154, 158
    strength, ix, 127, 159, 163
vorticity, 103, 104, 153

W

wind tunnel, 9, 32, 69, 70, 71, 97, 109, 125, 131, 132, 134, 142
work
    boundary, 87, 89, 92
    flow, 88, 89, 92
    shaft, 88, 89
Wright
    brothers, 9
    Flyer I (aircraft), 9
    Orville, 9
    Wilbur, 9

X

xenon, 55
XFLR-5, viii, 154, 156, 158, 194

Y

yaw, 3, 5, 6, 10, 12, 70, 123, 171, 189

ANSWERS TO SELECT PROBLEMS

1.1) a) 83,333 gal; 66.96 pass·mi/gal; 4.63% increase

   b) $208,333.33; 2.695 x 10^6 kWhr; $161,734.54; State of the art electric power sources have yet to reach the energy density of fossil fuels, making current technologies for electric power on long-haul aircraft impractical

1.3) Equilibrium

2.3) 0.0608 kg/m^3; 2.0591 kg/m^3

2.5) 0.1691 kg/m^3; 5,065 N; 31,634 N

2.8) 301,449 Pa; 30.8 mH$_2$O

3.2) a) 0.48 m/sec^2
   b) 2191 m/sec; 7.97 hr

3.3) a) 0.07272 kg/m^3
   b) 43,780 Pa
   c) 0.0609 kg/m^3
   d) 6.37 m/sec^2
   e) 20,080 m

3.7) a) 0.5874 kg/m^3
   b) 6,500 m
   c) 4,000 m
   d) 7,000 m

4.3) a) 68.6 kg/sec; 9,604 N
   b) 70.11 kg/sec; 52,582 N
   c) 0.4035 kg/m^3; 0.232 m^2; 42,978 N

4.5) a) 1.150 kg/m^3
   b) 38.1 kg/sec
   c) 305.8 K
   d) 103,271 Pa

4.11) a) 0.889 mm
   b) 36.5 mm

5.6) a) 138.4 Pa
   b) 4.5
   c) 8.65 N/m
   d) 0.490 m^2/sec

5.10) a) 0.25
   b) 6.4
   a) 0.868
   b) 0.280
   e) 4.79x10^6

UNITS, CONSTANTS, AND CONVERSIONS

Sea Level Standard Atmospheric Values

absolute alt., $h_{a,sl}$	$20{,}910\,kft$	$6{,}357\,km$	density, $\rho_{sl,std}$	$0.002377\,slug/ft^3$	$1.225\,kg/m^3$
pressure, $p_{sl,std}$	$2116\,psf$	$101{,}325\,Pa$	viscosity, $\mu_{sl,std}$	$3.7373\times10^{-7}\,slug/ft\cdot sec$	$1.7894\times10^{-5}\,kg/m\cdot sec$
temperature, $T_{sl,std}$	$519°R$	$288\,K$	gravity, g_0	$32.174\,ft/sec^2$	$9.8067\,m/sec^2$

Additional Useful Constants

R_u	$1545.4\,ft\cdot lb_f/lbmol\cdot°R$	$8314.3\,N\cdot m/kmol\cdot K$	$c_{v,air}$	$4290\,ft\cdot lb_f/slug\cdot°R$	$717.5\,N\cdot m/kg\cdot K$
R_{air}	$1716\,ft\cdot lb_f/slug\cdot°R$	$287\,N\cdot m/kg\cdot K$	$c_{p,air}$	$6006\,ft\cdot lb_f/slug\cdot°R$	$1005\,N\cdot m/kg\cdot K$
N_A	$6.022\times10^{23}\,mol^{-1}$		σ	$0.1714\times10^{-8}\,BTU/hr\cdot ft^2\cdot°R^4$	$5.670\times10^{-8}\,W/m^2\cdot K^4$
G	$3.450\times10^{-8}\,lb_f\cdot ft^2/slug$	$6.674\times10^{-11}\,N\cdot m^2/kg$			

Length Conversions

$1m=100\,cm=3.2808\,ft$	$1ft=12\,in=0.3048\,m$	$1mi=5280\,ft=1.609\,km$	$1km=0.6215\,mi$
$1in=2.54\,cm=0.0254\,m$	$1km=1000\,m=3281\,ft$	$1nmi=6076\,ft=1852\,m$	$1cm=0.394\,in$

Area Conversions

$1ft^2=144\,in^2=0.0929\,m^2$	$1in^2=0.006944\,ft^2$	$1m^2=10^4\,cm^2=10.76\,ft^2$	$1cm^2=0.155\,in^2$

Volume Conversions

$1US\,gal=0.1337\,ft^3$	$1US\,gal=3.785\,L$	$1ft^3=0.0283\,m^3$	$1m^3=35.335\,ft^3$
$1L=1\,dm^3=10^{-3}\,m^3$	$1cc=1\,cm^3=10^{-6}\,m^3$	$1ci=1\,in^3=16.4\,cm^3$	$1cm^3=0.061\,in^3$

Mass Conversions

$1kg=1000\,gm$	$1kg=2.2046\,lb_m$	$1kg=0.0685\,slug$	$1lb_m=0.4536\,kg$
$1lb_m=453.6\,gm$	$1lb_m=0.03108\,slug$	$1slug=32.174\,lb_m$	$1slug=14.594\,kg$
$1t=1000\,kg$	$1ton(US)=2000\,lb_m$	$1ton(UK)=2240\,lb_m$	

Density Conversions

$1kg/m^3=0.00194\,slug/ft^3$	$1kg/m^3=10^{-3}\,gm/cm^3$	$1slug/ft^3=515.38\,kg/m^3$

Force Conversions

$1lb_f=32.2\,lb_f\cdot ft/sec^2$	$1N=1\,kg\cdot m/sec^2$	$1N=0.225\,lb_f$	$1lb_f=4.448\,N$
$1lb_f=1\,slug\cdot ft/sec^2$	$1kip=1000\,lb_f$	$1dyn=1\,gm\cdot cm/sec^2$	$1N=10^5\,dyn$

Pressure Conversions

$1Pa=1.4504\times10^{-4}\,lb_f/in^2$	$1N/m^2=1\,Pa$	$1Pa=0.02088\,lb_f/ft^2$	$1bar=10^5\,Pa$
$1atm=2116\,lb_f/ft^2$	$1atm=101{,}325\,Pa$	$1atm=760\,mmHg$	$1atm=14.7\,lb_f/in^2$
$1inHg=3376.8\,Pa$	$1inHg=0.491\,lb_f/in^2$	$1inH_2O=5.2\,lb_f/ft^2$	$1inH_2O=248.8\,Pa$

Energy Conversions

$1J=1\,N\cdot m$	$1J=9.479\times10^{-4}\,BTU$	$1kWhr=3.6\,MJ$	$1BTU=1055\,J$
$1BTU=778.16\,ft\cdot lb_f$	$1BTU=252\,cal$		

Power Conversions

$1W=1\,J/sec=1\,VA$	$1W=3.412\,BTU/hr$	$1W=1.3405\times10^{-3}\,hp$	$1kW=1.3405\,hp$
$1BTU/hr=0.2161\,ft\cdot lb_f/sec$	$1BTU/hr=0.293\,W$	$1hp=550\,ft\cdot lb_f/sec$	$1hp=746\,W$

Velocity Conversions

$1mi/hr=1.46\,ft/sec$	$1mi/hr=1.609\,km/hr$	$1kt=1.687\,ft/sec$	$1m/sec=3.29\,ft/sec$

Temperature Conversions

$°F=\left(\frac{9}{5}\right)°C+32=°R-460$	$°C=\left(\frac{5}{9}\right)(°F-32)$	$°R=\left(\frac{9}{5}\right)K=°F+460$	$K=°C+273$